Web Mining: A Synergic Approach Resorting to Classifications and Clustering

RIVER PUBLISHERS SERIES IN INFORMATION SCIENCE AND TECHNOLOGY

Series Editors

K. C. CHEN
National Taiwan University
Taipei, Taiwan

SANDEEP SHUKLA
Virginia Tech
USA

CHRISTOPHE BOBDA
University of Arkansas
USA

The "River Publishers Series in Information Science and Technology" covers research which ushers the 21st Century into an Internet and multimedia era. Multimedia means the theory and application of filtering, coding, estimating, analyzing, detecting and recognizing, synthesizing, classifying, recording, and reproducing signals by digital and/or analog devices or techniques, while the scope of "signal" includes audio, video, speech, image, musical, multimedia, data/content, geophysical, sonar/radar, bio/medical, sensation, etc. Networking suggests transportation of such multimedia contents among nodes in communication and/or computer networks, to facilitate the ultimate Internet.

Theory, technologies, protocols and standards, applications/services, practice and implementation of wired/wireless networking are all within the scope of this series. Based on network and communication science, we further extend the scope for 21st Century life through the knowledge in robotics, machine learning, embedded systems, cognitive science, pattern recognition, quantum/biological/molecular computation and information processing, biology, ecology, social science and economics, user behaviors and interface, and applications to health and society advance.

Books published in the series include research monographs, edited volumes, handbooks and textbooks. The books provide professionals, researchers, educators, and advanced students in the field with an invaluable insight into the latest research and developments.

Topics covered in the series include, but are by no means restricted to the following:

- Communication/Computer Networking Technologies and Applications
- Queuing Theory
- Optimization
- Operation Research
- Stochastic Processes
- Information Theory
- Multimedia/Speech/Video Processing
- Computation and Information Processing
- Machine Intelligence
- Cognitive Science and Brian Science
- Embedded Systems
- Computer Architectures
- Reconfigurable Computing
- Cyber Security

For a list of other books in this series, www.riverpublishers.com

Web Mining: A Synergic Approach Resorting to Classifications and Clustering

V. S. Kumbhar
Department of Computer Science
Shivaji University
Kolhapur, India

K. S. Oza
Department of Computer Science
Shivaji University
Kolhapur, India

R. K. Kamat
Department of Computer Science
Shivaji University
Kolhapur, India

River Publishers

Routledge
Taylor & Francis Group

LONDON AND NEW YORK

Published 2016 by River Publishers
River Publishers
Alsbjergvej 10, 9260 Gistrup, Denmark
www.riverpublishers.com

Distributed exclusively by Routledge
4 Park Square, Milton Park, Abingdon, Oxon OX14 4RN
605 Third Avenue, New York, NY 10158

First published in paperback 2024

Web Mining: A Synergic Approach Resorting to Classifications and Clustering / by V. S. Kumbhar, K. S. Oza, R. K. Kamat.

Routledge is an imprint of the Taylor & Francis Group, an informa business

Publisher's Note
The publisher has gone to great lengths to ensure the quality of this reprint but points out that some imperfections in the original copies may be apparent.

While every effort is made to provide dependable information, the publisher, authors, and editors cannot be held responsible for any errors or omissions.

ISBN: 978-87-93379-83-1 (hbk)
ISBN: 978-87-7004-451-6 (pbk)
ISBN: 978-1-003-34003-4 (ebk)

DOI: 10.1201/9781003340034

Contents

Preface

The 21st century also regarded as the knowledge era has witnessed upsurge in the field of Information and Communication Technology (ICT). Especially in the last decade, there are many instances of increasingly more entities becoming online. The proliferation of the Internet addresses through IPV6, creation of newer domains online, social networking, and e-commerce are some of the forces behind the increasing online transactions. Paradigm shift of the traditional sectors to the online domain is no doubt a boon for the society. The ease and comfort with which the intended work is getting done have really fuelled online phenomena and, in the true sense, the notion of 'global village' is seen coming through.

However, from the scientific and technological point of view, this intensive online phenomena based on the ICT have poised several research problems. The sea of data being crunched is increasingly accelerated with great pace. This seems to pose the problems of its storage, processing and conversion in useful form. This has also given rise to many complementary developments in the form of parallel processing, which addresses the processing issues. Cloud Computing-based solutions are also being explored to sort out the storage and processing issue albeit with some cautions of privacy and security. One of the main issues the present research work is addressing is related to data mining. Main applications of different online domains develop an efficient and scalable technique, which may be able to handle different data types. Many researchers are also concentrating on this issue and try to develop a new and improved technology. Different methodology and various algorithms are being developed to address the woes of the big data. Thus, the data mining is one of the important technologies, which come up with great ideas taken from various disciplines such as statistics, artificial intelligence, machine learning and pattern recognition. Techniques such as sampling, estimation, and hypothesis testing are from statistics, searching algorithms, modeling, learning from artificial intelligence, etc. Data mining also receives support from other disciplines such as optimization, evolutionary computing, information theory, signal processing, visualization, and information retrieval. In the backdrop of the above scenario, the present book is an attempt to present the know-how pertaining to classification, clustering and web data mining.

The book is divided into four chapters. Chapter 1 presents introduction and prepares the potential reader providing with the requisite background of the subject. Chapter 2 depicts the prior research work undertaken by well-known research groups all over the world. Dataset creation is a formidable task in the domain of web mining. Chapter 3 takes the reader systematically through the process of dataset creation. Chapter 4 then reveals the nitty gritty of classification and clustering, which is the main aim of the book. It also presents glimpses of the work in the nutshell and divulges the future possible research directions pertaining to the web mining.

Dr. V. S. Kumbhar
Dr. K. S. Oza
Dr. R. K. Kamat

Acknowledgment

The present book is a scholarly outcome of support, motivation and peer review of our colleagues and friends. At the outset thanks are due to the authorities of the Shivaji University, Kolhapur for encouraging this research endeavour and providing infrastructural support for the same. Heartfelt thanks to Professor Devanand Shinde Vice-Chancellor, Shivaji University for his motivation in documenting our research in the form of book.

One of the authors Professor R. K. Kamat would like to acknowledge support received under the UGC SAP program which was decisive in showing this book the light of the day. Authors are grateful to Professor D. T. Shirke for his guidance during the course of the book. All the authors are thankful to their family members for the time they permitted in completion of the book.

Thanks are also due to Mr. Mark De Jongh, Junko Nakajima and the copy editors of the book for coming out in a very elegant style. Thanks to the anonymous reviewers for their constructive criticism which helped us in improving the book content.

Dr. V. S. Kumbhar
Dr. K. S. Oza
Dr. R. K. Kamat

List of Figures

List of Tables

List of Graphs

List of Abbreviations

ANN   Artificial Neural Network
CART   Classification and Regression Tree
CRM   Customer Relationship Management
CSS   Cascading Style Sheets
DGA   Dissolved Gas Analysis
DNA   Deoxyribonucleic Acid
DNS   Domain Name System
ECG   Electrocardiogram
EDM   Educational Data Mining
GDW   Geographic Data Warehouses
GGP   Genetic Programming Algorithm
GIS   Geographic Information Systems
GKD   Geographic Knowledge Discovery
GUI   Graphical User Interface
HR   Human Resource
HTML   Hyper Text Markup Language
HTTP   Hyper Text Transfer Protocol
ID3   Iterative Dichotomiser
IT   Information Technology
KDD   Knowledge Discovery in Databases
MLP   Multilayer Perceptron
MRI   Magnetic Resonance Imaging
OLAP   Online Analytical Processing
RBF   Radial Basis Function
RBI   Reactive Business Intelligence
ROI   Return On Investment
SEM   Strategic Enterprise Management
SEO   Search Engine Optimization
SMO   Sequential Minimal Optimization
SOM   Self-Organizing Map
SPECT Single Photon Emission Computed Tomography

SQL	Structured Query Language
SVM	Support Vector Machine
URL	Uniform Resource Locator
WWW	World Wide Web
XML	Extended Markup Language

1

Introduction

Abstract

This chapter describes the notion of Data Mining in depth. Data mining has been used traditionally for extracting the hidden, potential, useful and valuable information from very large amount of data, thanks to the state-of-the-art Data mining tools, which can potentially handle high dimensionality, heterogeneous and complex type data, including the so-called non-traditional data. Earlier such data was available in simple form and stored at a central place. Such data was used to mine by firing queries in order to find out the hidden patterns and in turn devising the knowledge from within. In current scenario, traditional techniques are unable to cater to the big multidimensional data since the size is too large and has been distributed at different places besides the issue of heterogeneity. Moreover, the real-world data is always dirty so there is need to convert it into quality data by using data preprocessing techniques. The data preprocessing includes data cleaning, data integration, data selection, and data transformation. After preprocessing, data mining techniques can be applied on such error-free data in order to extract the hidden knowledge. In addition, Data mining strategies include classification, clustering, association, prediction, estimation, etc. Classification techniques are used to classify the given datasets into two or more distinct classes, for example, decision tree, Naive Bayes, SVM, ANN etc. Clustering techniques are used to categorize the given data into a number of clusters so that inter-cluster data items are similar and intra-cluster data items are dissimilar, for example, k-means. Association rule finds the links in the data. The potential applications of data mining include financial data analysis, retail industry, telecommunication industry, biological data analysis and other scientific applications. Recent trends in data mining described in this chapter are visual data mining, distributed data mining, web mining, graph mining, etc. The issues related to classifications are data cleaning, relevance analysis, and data transformation and reduction. In the present chapter, we describe Weka, which is used for the analysis of

1

collected data. Weka is open source software developed at Waikato University at New Zealand, which the potential readers would enjoy in their applications pertinent to big data analytics as the one presented in this book.

1.1 Basic Notion of Data Mining

It is worthwhile at the outset to get to know the basic notion of data mining which forms the basis of web mining.

There are few definitions of data mining. Some of them are given as follows:

- Data mining is a technology that tries to blend different types of traditional data analysis methods with advanced and sophisticated algorithms for processing very large volume of data. It is the process of discovering hidden, valuable information in huge amount of homogeneous data [1].
- Gordon S. Linoff et al. defined data mining as a business process which explores huge amounts of data to find out useful, meaningful and valuable hidden patterns and rules [2].
- AB. M. Shawkat Ali, and others defined data mining as the art and science of extracting useful information which is hidden in large datasets [3].

Data mining is the search for new, nontrivial, useful, and valuable information in excessive datasets. Data mining is a process of discovering different methods, summaries, and derived values from a given collection of datasets [4]. "Data mining is the practice of automatically searching large stores of data to discover patterns and trends that go beyond simple analysis. Data mining uses sophisticated mathematical algorithms to segment the data and evaluate the probability of future events." [5]. The key properties of data mining are: (i) automatic discovery of patterns, (ii) prediction of likely outcomes, (iii) creation of actionable information, and (iv) focus on large datasets and databases [5].

1.2 Knowledge Discovery: The Very Rationale Behind Data Mining

The very rationale behind data mining is devising knowledge out of the sea of data. Jiawei Han and Michaeline Kamber stated, "data mining as it refers to extracting knowledge from large amounts of data". Data mining has been referred with other names as knowledge discovery in databases (KDD), knowledge extraction, data analysis, pattern analysis, data archeology, data dredging, business intelligence, etc. Data mining is one of the important steps in the Knowledge Discovery Process (KDD) [6].

KDD process has the following different steps:

- **Data cleaning**: In data cleaning process, noise is detected and removed from the inconsistent data, so that data becomes free of errors.
- **Data integration**: In this step, data is available at different locations and it is combined at a central location to perform mining on it.
- **Data selection**: The data is stored into the databases. It is necessary to retrieve the data which is relevant for the analysis.
- **Data transformation**: In this step, data is transformed into suitable forms so that mining can be performed by using aggregation operations, and summary operations.
- **Data mining**: Data mining is used to extract valuable, hidden information by applying intelligent methods.
- **Pattern evaluation**: After applying data mining on the data, different types of patterns are generated. All these patterns are not useful for us. Therefore, we try to identify actual interesting patterns, which are representing knowledge based on some interestingness measures.
- **Knowledge presentation**: This is the last step where user gets mined knowledge for that various visualization and knowledge representation techniques used.

First four steps are used for preparing the data suitable for the mining task.

These first four steps of KDD process are called data preprocessing steps. In data preprocessing, rough, raw, dirty data is converted into quality data which is free of error [6] (Figure 1.1).

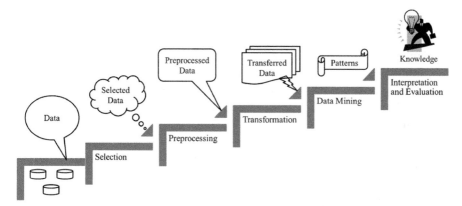

Figure 1.1   Different stages in the KDD process.

Data mining techniques are potentially used to find useful and novel patterns that might be unknown to us. Not all tasks of searching information are considered as data mining tasks. Data mining is one of the key activities in the KDD process where raw data is converted into knowledge. The input data can be made available in the form of relational tables, spreadsheets, word file, flat files, etc., and may be stored in centralized data repositories or be distributed across multiple sites. The data preprocessing is used to convert the raw input data into quality data for analysis purpose. For this purpose, different techniques are used, such as cleaning data to remove noise and duplicate observations, and selecting records and features that are relevant to the data mining task. Data preprocessing is one of the tedious and time-consuming job in the KDD process because there are multiple ways of data collection and data storage [1]. In the early days of data mining, data was rare. Owing to the rapid development of the hardware and software tools, it is possible to collect and analyze large amount of data. Data mining techniques are useful for such type of data [2].

1.3 Challenges in the Development of Data Mining

In the wake of increasing heterogeneity of data, immense challenges are emerging. Some of them are as follows.

1.3.1 Scalability

Nowadays, generation and collection of data is a routine task. Owing to this, the size of the datasets becomes common, i.e. gigabytes, terabytes or even petabytes. If such huge datasets are handled by the data mining algorithms, then they must be scalable. Most of the data mining algorithms use special search strategies to handle problems of exponential search. Scalability is to access individual records efficiently and effectively, and it requires implementing proper data structure and algorithm. For instance, intelligent algorithms may be essential when processing such a huge dataset that cannot fit into computer's main memory. To achieve scalability sampling or parallel and distributed algorithms, some techniques may be used [1].

1.3.2 High Dimensionality

Few decades ago, datasets have only few attributes. Nowadays, it is common that datasets have hundreds or thousands of attributes. Owing to fast evaluation in microarray technology, gene expression data in bioinformatics involves thousands of attributes. Spatial and temporal datasets are also having high dimensionality, for example, dataset of temperatures at various locations. As temperatures are taken for specific period repeatedly, dimensions increase in proportion to temperature measure. Traditional data analysis techniques were capable to handle less-dimensional data and therefore not useful for high-dimensional data. For any data analysis algorithm, if dimensionality of data increases, computational complexity also increases [1].

1.3.3 Heterogeneous and Complex Data

Traditional data analysis techniques works well for the same type and continuous or categorical type attributes. Data mining playing a vital role in business, science, medicine, and other fields has developed, and hence there is imminent need to develop techniques that can handle heterogeneous attributes. Recently, data come up with very complex nature. For example, Web page data, semi-structured text and hyperlink data, audio and video data, unstructured data, DNA data, and climate data that consist of time series measurements (temperatures, pressure, etc.). Nowadays, researchers are taking efforts to handle such complex objects in order to find the interesting patterns that are hidden in it [1].

1.3.4 Data Ownership and Distribution

Generally, the data used for processing is stored in centralized server. Sometimes, the data required for the analysis purpose is not stored in a particular location or it is not of one organization. Geographically data is distributed among the resources which have multiple entities. This leads to the development of new techniques called distributed data mining. Distributed data mining faces challenges like:

(i) how to reduce the actual communication for performing distributed computation;
(ii) how to combine the data mining results, which are obtained from various sources; and
(iii) how to handle different types of data privacy and data security issues [1].

1.3.5 Non-Traditional Analysis

In statistical processing, first, a hypothesis is proposed, then the experiment is designed to collect data, and lastly, collected data is analyzed according to the stated hypothesis. This traditional method is labor intensive. Frequently, the current data analysis tasks need generation and evaluation of hypotheses. This process of hypothesis generation and evaluation is automated by data mining techniques. In data mining, the datasets analyzed are not the result of a designed experiment and they represent data samples instead of random samples. The datasets also involve non-traditional data and data distributions [1].

1.4 Importance of Data Mining

Nowadays, everywhere, a large amount of data is generated. The challenge is to store such data and handle them for processing. It is necessary to convert such a large amount of data into useful information and knowledge. Owing to this, data mining attracts the attention of people working in society and information industry. The acquired information and knowledge is used in a variety of applications, such as in science and engineering, customer retention, fraud detection, and market analysis.

In general, a result of progression of information technology is data. The database system industry already played a very important role in the development of the functionalities such as collection and creation of data, management of data, and advanced data analysis. The management of data includes data storage and data retrieval, and database transaction processing. The advanced data analysis includes data warehousing and data mining. Today's advanced database technologies are possible only due to the consistent development in the existing and old one such as data collection and data creation mechanism, query and transaction processing, and data storage and retrieval mechanisms.

The progress has been made from file-processing systems to sophisticated and powerful database systems in the database and information technology since 1960. There is continuous research and development in the database systems. As a result of this since the 1970s, database systems have been developed from hierarchical and network database systems to relational database systems, indexing and accessing methods and data modeling tools. The user can handle the data by using query language, user interface, etc. (Figure 1.2).

Since the 1960s, traditional file-processing system will evaluate to powerful and sophisticated database and information technology. Since 1970, traditional hierarchical and network models are shifted to relational database models, data modeling tools, and indexing and accessing methods. Also, a user can access the data using query language, user-friendly interfaces, optimized

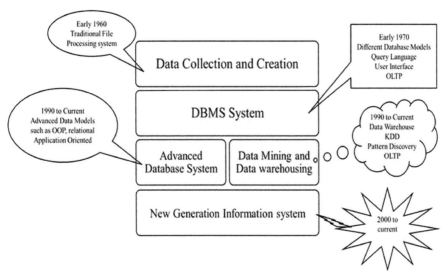

Figure 1.2 Historical aspects of data mining.

query-processing techniques, and advanced, secure transaction management. OLAP technology can be efficiently used for storage, retrieval, and handling huge amount of data.

Since the mid-1980s, the database technology has evolved to rational technology. There is an enhancement of research and development activities on upcoming new and powerful sophisticated database systems. The result is advanced data models like extended-relational, object-oriented, object-relational, and deductive models. The application-oriented database systems have flourished which includes spatial, temporal, stream, sensor, multimedia, science and engineering databases, etc. The different types of issues can be handled very extensively by taking a lot of efforts such as sharing, distribution, and diversification of data. At the same time, data is available at distributed locations and in variety of forms such as the World Wide Web (WWW). Therefore, the database and information industry deeply think about how to manage such type of huge amount of heterogeneous data and extract the valuable information and knowledge from it. The web mining is the recent technique used to extract information and knowledge from large amount of differently distributed data sources. Data may be semi-structured or unstructured.

In the last 30 years, there has been a continuous and amazing progress in the field of computer hardware and it leads to the development of powerful computers, data collection devices, and different storage technologies. Owing to this technology, it is possible in database and industry to make a large number of repositories for the data analysis, information retrieval, and transaction management.

Now, it is possible to store data in different categories of databases and information repositories. For the same, data warehouse is used where data may be stored from multiple heterogeneous data sources. This data is helpful for making proper decisions. Data warehouse technology consists of data cleaning, data integration, and on-line analytical processing (OLAP). In OLAP, different types of functionalities are used such as summarization and aggregation, and the information can be viewed from different angles. Also, OLAP supports analysis of multidimensional data. In addition to this, tools for data analysis are required to perform in-depth analysis of data such as clustering, classification, and data characterization. The size of data growing day-by-day and present databases and data warehouses are unable to handle such a huge amount of data, for example, data streams and WWW where data flows in and out. Such types of stream data come in the application like sensor networks, telecommunications and video surveillance. The challenging task is that of handling such stream data effectively and efficiently, in order to process and extract the hidden, useful, and valuable patterns.

There is heavy flow of data. Powerful and sophisticated data analysis tools are required to handle this huge data. The situation is just like 'a *data rich but information poor* situation'. It is very difficult to collect, store, and handle such fast-growing, tremendous amount of data. For processing, we require powerful tools, so that we can easily extract useful information and knowledge. Actually the important decisions solely depend on the information-rich data. For handling tremendous volume of data stored in large repositories, data mining technology is used which will definitely convert data tombs into valuable knowledge [6].

1.5 Classification of Data Mining Systems

Data mining is an interdisciplinary field, which includes database systems, visualization, statistics, machine learning, algorithms, pattern recognition, etc. This is shown in figure. The concepts or techniques are applied from other subjects such as rough set theory, neural network, fuzzy logic, visualization techniques, parallel and distributed computing, etc., on the data mining for properly finding the interesting patterns in the datasets. Depending on the data mining applications, data mining systems may try to use the concepts or techniques not only from science-related fields such as information retrieval, pattern recognition, spatial data analysis, image processing and analysis, signal processing, computer graphics, web technology, and bioinformatics but also from commercial fields such as economics and business (Figure 1.3).

Data mining technology is used in a number of fields, due to the diversity of disciplines helping data mining systems to generate a bulky amount of data and store it in data warehouse. The large number of data mining systems is expected in data mining research. Therefore, it is necessary to classify the data mining systems, so that users can select the best one for their work. According to the different criteria, data mining systems can be classified as follows.

1.5.1 The Databases Mined

The data mining systems have various databases. From these databases, knowledge is to be mined. According to the various criteria such as data models or types of data, the database systems are classified. Each and every database system may require its specific data mining technique. If the criteria

Figure 1.3 Interdisciplinary data mining.

are data models, then we may have different mining systems such as relational, transactional, object-relational, etc., and if the criteria are types of data, then we may have a spatial, temporal, time-series, text, stream data, and multimedia or WWW mining system.

1.5.2 The Knowledge Mined

Depending on the data mining functionalities, the knowledge such as discrimination, characterization, association and correlation analysis, prediction, classification, clustering, outlier analysis, etc., can be mined. These kinds of knowledge can be mined using details of data, i.e. granularity of data. The knowledge can be mined at a high level of abstraction, at a rough data level of abstraction, or at multiples levels of abstraction. For achieving these types of mining at different levels of abstraction, advanced data mining systems are required so that anyone can extract the knowledge from such a huge amount of data. There are other applications where we try to find the unexpected values in the datasets called outliers. At first, we must detect the outliers and then remove them from the datasets. There are so many techniques used to detect outliers in the datasets, such as clustering, classification, prediction, correlation analysis, etc.

1.5.3 The Techniques Utilized

Categories of data mining system are based on the data mining techniques used. It also uses multiple data mining techniques that collect the merits of some individual approaches.

1.5.4 The Application Adopted

According to the application, data mining systems are categorized, for example, stock market, e-mail, DNA, telecommunications, finance, etc. [6].

1.6 Generic Architecture of Data Mining System

The components of data mining architecture are:

- **Different Data Repository**

It is a collection of different data repositories such as various databases, flat file system, data warehouse, spread sheet, or some other information repositories. From this heterogeneous source, data will be collected and data pre-processing technique will be applied for further process.

- **Database or data warehouse server**

According to the procession task or data mining problem, relevant data is fetched.

- **Knowledge Base**

This step is used to search and identify the usefulness of the discovered pattern. Knowledge base represents the concept pyramid according to the level of abstraction. User opinion is one form of knowledge helpful for identifying pattern's interestingness. Other kinds of domain knowledge are additional interestingness constraints or threshold and metadata.

- **Data Mining Engine**

Data mining engine consists of a collection of different models used for classification, clustering, characterization, association, correlation, and evolution analysis.

- **Pattern Evaluation Module**

This component generally useful for interestingness measures and interrelates data mining modules. It is used to focus search toward user's interest. It filters out patterns by using some threshold value. This model combines with data mining implementation methods. For efficiency purpose, one can use constraints which will be push deep into the mining process. Herein, constraint represents interestingness of the pattern. By using this method, search space will be reduced.

- **User Interface**

This component works as an interface between the end user and the data mining system. Interaction can take place by using data mining query language. The end user can provide required information to the mining process, which will focus on search and increase the performance of data mining process to yield better results. This is helpful to locate database and data warehouse schemas, evaluate mined patterns and data structures, and visualize the patterns in different forms to the user [6] (Figure 1.4).

It is not an easy task to measure the effectiveness or usefulness of a data mining approach for extracting valuable patterns from the huge amount of data. Different metrics are used for different techniques. Return on investment (ROI) is used as a measure for effectiveness. ROI tries to optimize the process and gives several types of benefits and it is challenging to determine. It could be measured as increasing the profit or decreasing the cost [7].

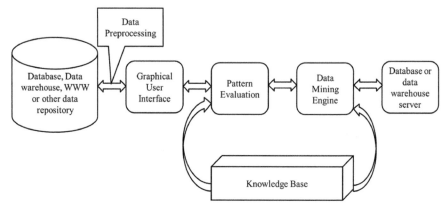

Figure 1.4   Data mining architecture.

1.7 Major Issues in Data Mining

1.7.1 Mining Methodology and User Interaction Issues

It is concerned with different kinds of knowledge mined, mining at multiple level and granularities, and try to use if domain knowledge, and knowledge visualization.

- **Mining efficient kinds of knowledge in database**: The data is stored into the databases. The same data can be used by different customers, because they are interested in different kinds of knowledge. For this, data mining technology supports knowledge mining by using different techniques such as classification, clustering, outlier analysis, prediction, etc. All these techniques may use the same databases in different ways and display the output as per the user's needs.
- **Interactive mining of knowledge at multiple levels of abstraction**: Before processing data, sometimes it is very difficult to know exactly what is to be discovered in a given database. For this, there must be an interaction between the user and the database systems. If databases have a tremendous amount of data, suitable sampling techniques can be applied on the selected data. Owing to interactive mining, it is possible for the user to focus on different search patterns. To apply different types of OLAP operations on the datasets, such as drilling down, roll up, pivoting, slice, dice for mining knowledge. In this way, the user can interact with the data mining system to view data through different angles and discover patterns at multiple granularities and from different angles.
- **Incorporation of background knowledge**: To guide and help for discovery process, information regarding the domain or background knowledge

is must. It is possible to express the discovered patterns and levels of abstractions. The domain knowledge helps to speed up the process of data mining.

- **Data mining query languages and ad hoc data mining**: The user can fire ad hoc queries on relational query languages for data retrieval. Similarly, high-level data mining query languages are used to describe ad hoc data mining task by taking the relevant datasets for analysis, the knowledge to be mined, etc. Such a language should be combined with a database or data warehouse query language and improved for efficient data mining.
- **Presentation and visualization of data mining results**: The data mining techniques are applied on datasets to extract the valuable patterns and knowledge. Once these patterns and knowledge are extracted, it is a challenging task to express it and represent it in visual forms, so that the user can easily understand and use as it is. For this, sophisticated knowledge representation techniques are required such as trees, graphs, charts, curves, etc.
- **Handling noisy or incomplete data**: The data stored in a database or data warehouse may contain noise, or data is incomplete, or data is inconsistent. Sometimes, such raw data processed by the data mining system may lead to wrong results. The process gets confused and overfits the data in the model. Owing to this, the discovered patterns are poor and accuracy gets decreased. Data cleaning and analysis methods are required to detect and remove noise from such a dirty data. At the same time, outlier analysis methods are also helpful for detecting and removing the outlier from the rough data.
- **Pattern Evaluation—the interestingness problem**: After processing datasets, data mining system can discover a number of patterns. Many of these discovered patterns are not useful to the user or the user is not interested in it. As per the expectations of the user, the data mining system should generate and discover the patterns in the datasets. A few metrics are used to find out the interestingness discovered patterns in the datasets.

1.7.2 Performance Issues

This is related to efficiency, scalability, and parallelization of data mining algorithms.

- **Efficiency and scalability of data mining algorithms**: The data mining algorithms must be efficient and scalable to find out useful information from an excess amount of data. When the size of datasets is very large, the time complexity of the data mining algorithms must be predictable and acceptable.

- **Parallel, distributed, and incremental mining algorithms**: The size of the databases are very large, sometimes the data may be available at different isolated locations, and time complexity of some data mining algorithms is a key issue for handling such types of diverse data. The data of this type can be divided into a number of parts and processed in parallel, and then the results are merged. The cost of such algorithms is high. That is why there is a need to develop incremental data mining algorithms, which record data updates.

1.7.3 Issues Relating to the Diversity of Database Types

- **Handling of relational and complex types of data**: Mostly, people use relational databases and data warehouse for the processing. For such type of data, it is necessary to develop effective and efficient data mining systems. However, some databases may exist with complex data objects, spatial and temporal data, and hypertext and multimedia data. One data mining system is unable to find out patterns in the diverse types of databases. Hence, for such data, specific types of data mining techniques are required to be developed. Therefore, there may be different data mining techniques for different kinds of data.
- **Mining information from heterogeneous databases and global information systems**: Through Internet, it is possible to collect the data from many sources. When data is collected from so many resources, the size of the data becomes more and more big. This data may be available at distributed locations and it may be of homogeneous or heterogeneous type. It is so much difficult and a challenging task to discover the interesting patterns from such structured, semi-structured, or unstructured data. Web mining is concerned with finding the knowledge about web contents, web usage, and web structures. It is an upcoming field in data mining to extract the knowledge from data having no unique format [6].

1.8 Data Mining Strategies

There are different types of data mining strategies depending on the range of the application domain and the interests of the users. The following is the list of data mining strategies:

1.8.1 Classification

Classification problem can be stated as identifying an object which is belonging to a given class. All classification methods come under supervised learning. For a set of objects with given attribute values, the aim of the classification task is to develop the model by discovering the influences of the input attributes that decide whether the object belongs to a particular class. The model is used to predict the class of a new object. A class has only discrete values, namely Class 1, 2, 3, etc. Therefore, many classification algorithms have been developed for categorical data only. Classification techniques play an important role in the fields of bioinformatics, medical diagnosis, fraud detection, loan risk prediction, text classification, and engineering fault detection. The list of classification techniques are: Decision tree, Bayesian, rule-based, SVM, ANN, etc. [3].

Decision tree: The Decision tree is one of the most popular and powerful classification algorithms used in data mining. The decision tree represents rules. Rules can be expressed so that anybody can understand them or even directly used in database access language like SQL so that records falling into a particular category may be retrieved. Decision tree is a flowchart-like tree structure, where each internal node denotes a test on an attribute, each branch represents an outcome of the test, each leaf node indicates the value of the target attribute of examples, and the topmost node in a tree is the root node.

The decision tree used for classification:

- Given an instance, X, for which the associated class label is unknown.
- The attribute values of the instance are tested against the decision tree.
- A path is traced from the root to a leaf node, which holds the class prediction for that instance.

The decision tree algorithms have been used for classification in many areas, such as medicine, manufacturing and production, financial analysis, astronomy, and molecular biology.

Advantages of decision tree are as follows:

- it requires little data preparation,
- it is able to handle both categorical and numerical data,
- it is simple to understand and interpret,
- it generates a model that can be statistically validated,
- the construction of decision tree classifiers does not require parameter setting,
- decision trees can handle high-dimensional data,

- it performs well with large data in short time,
- the learning and classification steps of decision tree induction are simple and fast,
- accuracy is comparable to other classification techniques for many simple datasets,
- convertible to simple and easy to understand classification rules, and
- easy to interpret for small-sized trees.

The basic algorithms for tree constructions are ID3, C4.5, and CART. All these are greedy algorithms.

- The tree is constructed in a top-down recursive divide-and-conquer manner.
- At the beginning, all the training examples are at the root.
- Attributes are categorical.
- Examples are partitioned recursively into smaller subsets as the tree is being built based on selected attributes.
- Test attributes are selected on the basis of a statistical measure.

There are three popular attribute selection measures:

- Information gain: This measure is used to select among the candidate attributes at each step while growing the tree.
- Gain ratio: It is the ratio of information gain and entropy.
- Gini index: Gini index is an impurity-based criterion that measures the divergences between the probability distributions of the target attribute's values [6].

1.8.2 Association

It comes under unsupervised learning. The association task includes finding the links in data—things that go together or things that follow one another. Given a transaction database which describes a set of items, association technique finds all the items that most of the customer frequently purchase together in one transaction. For example, in association rule mining for a retail chain, data are collected using a device like bar-code scanners. Retail chain database consists of a large number of transaction records. Each record displays the list of all items purchased by a customer in a single transaction. Owner or managers try to know if a particular group of items is consistently purchased together by the customers. These people try to use such information for adjusting places of items, for cross-selling, for promotions, for design of catalog, and for identifying the customer group based on purchasing habits [3].

1.8.3 Clustering

It comes under unsupervised learning. Clustering is a process of grouping a given set of unlabelled patterns, so that the similarity of patterns within the same cluster is maximized, while the similarity of patterns between clusters is minimized. For successful clustering of algorithm, it is necessary to identify the attributes that group the instances into a number of distinct classes from those attributes which merely show clutter in the data [3].

k-means clustering: k-Means clustering intends to partition n objects into k clusters, in which each object belongs to the cluster with the nearest mean. This method produces exactly k different clusters of greatest possible distinction. The best number of clusters k leading to the greatest separation (distance) is not known a priori and must be computed from the data. The objective of k-Means clustering is to minimize the total intra-cluster variance, or the squared error function.

1.8.3.1 k-Means algorithm

Step 1: Clusters the data into k groups where k is predefined.

Step 2: Select k points at random as cluster centers.

Step 3: Assign objects to their closest cluster center according to the Euclidean distance function.

Step 4: Calculate the centroid or mean of all objects in each cluster.

Step 5: Repeat steps 2, 3 and 4 until the same points are assigned to each cluster in consecutive rounds.

k-Means is relatively an efficient method. However, we need to specify the number of clusters, in advance, and the final results are sensitive to initialization and often terminate at a local optimum. Unfortunately, there is no global theoretical method to find the optimal number of clusters. A practical approach is to compare the outcomes of multiple runs with different k and choose the best one based on a predefined criterion. In general, a large k probably decreases the error but increases the risk of overfitting.

Strength of k-means method:

- Relatively efficient: $O(tkn)$, where n is # objects, k is # clusters, and t is # iterations. Normally, k, $t << n$
- Often terminates at a local optimum. The global optimum may be found using techniques such as deterministic annealing and genetic algorithms.

Weakness of k-means method:

- Applicable only when mean is defined, then what about categorical data?
- Need to specify k, the number of clusters, in advance

- Unable to handle noisy data and outliers
- Not suitable to discover clusters with non-convex shapes.

Applications of k-means methods are:

- Optical character recognition
- Biometrics
- Diagnostic systems
- Military applications.

1.8.4 Estimation

It also comes under supervised learning. In estimation, we have to assess the numeric value of a target variable based on other inputs. In classification, the output attribute is categorical, but in estimation, the output attribute is numerical. The estimation method produces the output as an infinite continuum of classes, but finite number of classes produced by the classification methods. Estimation algorithms works with the entire set of variables including the target variable during the training phase. Then, at the time of model application to a new example, it works without any target variable to estimate its value using the training knowledge. Estimation methods are used in daily temperature forecasts, monthly rainfall measures, stock prices, and trading [3].

1.9 Data Mining: Ever Increasing Range of Applications

1.9.1 Games

Nowadays, data mining is used in the field of game, for example, chess. Since the early 1960s, with the availability of oracles for certain combinatorial games with any beginning configuration, small-board dots-and-boxes, small-board-hex, and certain endgames in chess, dots-and-boxes, and hex, a new area for data mining has been opened. This is the extraction of human-usable strategies from these oracles. Current pattern recognition approaches do not seem to fully acquire the high level of abstraction required to be applied successfully. Instead, extensive experimentation with the table bases answers to well-designed problems, and with knowledge of prior art is used to find insightful patterns. Berlekamp in dots-and-boxes, etc., and John Nunn in chess endgames are examples of researchers spending time for doing this work.

1.9.2 Business

In business, static data is stored in data warehouse databases. This historical data is analyzed by using different and sound data mining techniques. The objective is to find out the hidden, valuable, and most useful patterns and trends in it. Data mining algorithms use advanced pattern recognition algorithms for analysis of tremendous amount of data to discover previously unknown strategic business information. For example, data mining is used in business for market analysis, finding the actual reason for manufacturing problems, to analyze the behavior of the customer, to attract more customers by giving discounts on purchasing the products, and to find out the class of customers.

- Nowadays, the rough data is being collected and stored by various organizations and companies with very high rate, all over the world, for example, Walmart processes approximately more than 20 million transactions per day. This collected information is kept in a centralized database, but such information is useless because it is not analyzed by any data mining software. If experts take this Walmart's transactional data and analyze by using data mining techniques, then they will be able to find trends in sales and trends in customer's purchase habits, developing the marketing strategies.

- Data about the user's behavior is being collected when customers use different types of cards at the time of transactions, for example, credit card at the time of billing and warranty card at the time of product purchased. Companies like Google, Facebook, etc., store the information of the users and, due to this, many people get disturbed and their privacy is violated. Other way, the personal information of the users may be useful for some task. For example, for a particular bazaar, every day, a number of customers are visiting for purchasing product and their personal data is also collected. By using this personal data, it is possible to find their purchasing habits, class of the customer, etc.

- Data mining also contributes to the field of customer relationship management (CRM) applications, to contact the customer frequently for collecting their opinions, their satisfactions, their likelihood of responding to an offer, etc. Advanced methods may be used to optimize resources, so that someone predicts to which channel and to which offer an individual is most likely to respond. In addition to this, sophisticated applications could be used to automate mailing. Once the results from data mining are determined, this "sophisticated application" can automatically send either an e-mail or a regular mail. Finally, without providing any offer, many people will take some action, and if an offer is given to the customer, then an increase in customer's response is determined. The technique of data clustering is used to automatically discover the groups within a customer dataset.

- Businesses employing data mining may see a ROI, but they also recognize that the number of predictive models can quickly become very large. For example, rather than using a single model to predict how many customers will churn, the business may choose to build a new separate innovative model for each region and customer type. Some businesses turn to more automated data mining methodologies where a large number of models are required to be maintained.
- Human resources (HR) departments make use of data mining to identify the different characteristics of their most successful employees. Different types of facilities and incentives are provided to the employee who is helping to satisfy the company's goal properly. Strategic Enterprise Management (SEM) applications help a company to achieve various types of corporate-level goals by translating it into operational decisions, such as planning of the production
- Data mining techniques are also used in retail sales known as market basket analysis. If an organization stores the records of customers who are purchasing different items, a data mining system helps to find the different patterns. While purchasing one item, the customer may want to purchase another item in connection with the first one. From this, the organization can find the association between different items and arrange the items in such a way that these items are easily available to the customers, so that the organization increases the profit margin, for example, the customer purchasing tooth brush may like to purchase tooth paste also, so arrangement of these products can be made in a such way that they are very close to each other. By using market basket analysis, it is also possible to identify the purchase patterns of the customer. The user data analyzed by companies is used to predict future buying trends.
- For business applications, data mining can be integrated into a complex modeling and decision-making process. Reactive business intelligence (RBI) gives an approach which integrates data mining techniques, modeling techniques, and interactive visualization techniques into a discovery.

1.9.3 Science and Engineering

Recently, data mining has been soundly used in diverse areas of science and engineering, such as genetics, medicine, bioinformatics, education, weather forecasting, and electrical power engineering.

- In human genetics, sequence mining helps us for understanding the mapping relationship between the inter-individual variations in human

DNA sequence and the variability in disease susceptibility. Simply, it finds out the changes in an individual's DNA sequence causing the risks of developing common diseases, which is very important for improvement diagnosis and preventing methods, and also useful for treatment, for example, multifactor dimensionality reduction.

- In electrical power engineering area, data mining technology has been used for monitoring the condition of high-voltage electrical equipments. The purpose is to find valuable information on the status of the insulation. One of the data clustering techniques, self-organizing map (SOM), has been applied to detect abnormal conditions and it tries to find the nature of the abnormalities.
- Dissolved gas analysis (DGA) has been available for many years where data mining methods have been used in power transformers. SOM methods have been applied for analyzing generated data and determining the trends in it.
- In recent years, peoples are interested to make use of data mining to investigate scientific questions in educational research domain called educational data mining (EDM). It is defined as the area of scientific inquiry centered on the development of methods for making discoveries within the unique kinds of data that comes from educational settings, and using those methods to better understand students and the settings which they learn in, for example, in mining data about how students choose to use educational software, and at the same time, data is considered at the keystroke level, answer level, session level, student level, classroom level, and school level. In the study of educational data mining, various parameters are playing an important role like time, sequence, etc. [8].

1.9.4 Human Rights

Data mining of government records—particularly records related to courts and prisons—enables the discovery of human rights violations in connection with generation and publication of invalid or fraudulent legal records by various government agencies.

1.9.5 Medical Data Mining

Daily medical sector generates very large amounts of heterogeneous data. For example, medical data may contain MRI images, SPECT images, ECG signals, blood and urine analysis report, clinical information like temperature, cholesterol level in the blood, etc., as well as the physician's interpretation.

Those who handle such a type of data find the gap between data collection and data comprehension. For this purpose, experts use computerized techniques. This task is devoted to the medical data mining field. For diagnosis purpose, the images are referred frequently, so it is a challenging task to extract the information from such huge amounts of image data. To resolve this problem, it is necessary to develop methods for efficient mining in databases of images. In addition to this, security and confidentiality of data are also important. Moreover, the physician's interpretation of images, signals, or other technical data is written in unstructured English, which is very difficult to mine.

1.9.6 Spatial Data Mining

It is the application of data mining methods to spatial data. The main objective of spatial data mining is to find patterns in data with respect to geography. Thus far, data mining and Geographic Information Systems (GIS) have existed as two separate technologies, each with its own methods, traditions, and approaches to visualization and data analysis. The immense explosion in geographically referenced data due to developments in IT, digital mapping, remote sensing, and the global diffusion of GIS emphasizes the importance of developing data-driven inductive approaches to geographical analysis and modeling. Data mining is beneficial for GIS-based applied decision-making. Nowadays, the task of integrating these two technologies has become very important, especially as various public and private sector organizations possessing vast amount of databases with thematic and geographically referenced data begin to realize the huge potential of the information contained therein. Among those organizations are:

- Public health services searching for explanations of disease clustering
- Environmental agencies assessing the impact of changing land-use patterns on climate change
- Geo-marketing companies doing customer partition based on spatial location.

1.9.7 Challenges in Spatial Mining

Geospatial data repositories have very large size. The existing GIS datasets are generally divided into feature and attribute components that are conventionally archived in hybrid data management systems. These are the range and diversity of geographic data formats, which present unique challenges. The digital geographic data revolution is creating new types of data formats, which are different from the traditional "vector" and "raster" formats. Geographic data repositories include ill-structured data, such as geo-images and geo-referenced multi-media.

There are lot of research challenges in geographic knowledge discovery and data mining. The following is the list of emerging research topics in the field:

- Developing and supporting geographic data warehouses (GDW'): Spatial data to be collected from various geographical locations by using different attributes. Apply preprocessing techniques of the data mining so that data gets converted into quality data or becomes free of error. To create an integrated GDW requires solving different issues related to spatial and temporal data interoperability—including semantic differences, referencing systems, geometry, accuracy, and position.
- Better spatiotemporal representations in geographic knowledge discovery: Generally, the current geographic knowledge discovery (GKD) methods use very simple representations of geographic objects and spatial relationships. Geographic data mining methods should find out or search for more complex geographic objects (i.e., lines, polygons, and surfaces) and relationships (i.e., non-Euclidean distances, direction). Further, the time dimension needs to be fully integrated into these geographic representations and relationships.
- Geographic knowledge discovery using diverse data types: GKD method uses traditional raster and vector methods for spatial and temporal data mining. Now, GKD methods should be developed that can handle diverse types of data, which includes imagery and geo-referenced multimedia data and dynamic data types such as video streams, animation clips, etc.

1.9.8 Temporal Data Mining

Data may contain attributes which are generated and recorded at different time slots. Herein, to find the meaningful relationships in the data, this may require considering the temporal order of the attributes. A temporal relationship may indicate a causal relationship, or simply an association between these attributes.

1.9.9 Sensor Data Mining

Wireless sensor networks can be used for collection of data for spatial data mining for a variety of applications such as air pollution monitoring.

A characteristic of such networks is that nearby sensor nodes monitoring an environmental feature typically noted similar values. This kind of data redundancy due to the spatial correlation between sensor observations inspires the techniques for in-network data aggregation and mining. By measuring the spatial correlation between data sampled by different sensors, a wide class of specialized algorithms can be used to develop more efficient spatial data mining algorithms.

1.9.10 Visual Data Mining

Traditionally, the music data is in analog form. Now, the music data is in the form of digital data. Owing to this, very large datasets have been generated, collected, and stored. To apply visual data mining techniques for discovering statistical patterns, trends, and information, this is hidden in data for building predictive patterns. It is observed that visual data mining is faster and much more intuitive than the traditional data mining. Computer vision may have collection of animated clips, real recordings of the different situations generating the visual data. This visual data can be mined for finding the hidden data patterns and uses it for further research.

1.9.11 Music Data Mining

Everyone is interested to listen to music and songs in different languages and themes. Music plays a vital role in everyone's life. With digitization process, such songs and music are stored in the digital form and collections of large data are formed. This very large digital data is used by music enthusiasts. This collection is available not only in the form of audio, video, and compact disk but also on the hard drive and on the web. It is difficult to us to keep track of music and the relations between different songs manually. For this, we require data mining and machine learning techniques to find the patterns in it.

1.9.12 Pattern Mining

"Pattern mining" is a data mining method that involves finding existing patterns in data. In this context, patterns often means association rules. The original motivation for searching association rules came from the desire to analyze supermarket transaction data, that is, to examine customer behavior in terms of the purchased products. For example, an association rules "onion $\Rightarrow$ potato (80%)" states that four out of five customers that bought onion also bought potato. Pattern Mining includes new areas such a Music Information Retrieval (MIR), where patterns seen both in the temporal and in the non-temporal domains are imported to classical knowledge discovery search methods.

1.9.13 Subject-based Data Mining

"Subject-based data mining" is a data mining method involving the search for association between individuals in data.

1.9.14 Knowledge Grid

Knowledge discovery "On the Grid" generally refers to conducting knowledge discovery in an open environment using grid computing concepts, allowing users to integrate data from various online data sources, as well as make use of remote resources, for executing their data mining tasks. An example is the work conducted by researchers at the University of Calabria, who developed Knowledge Grid architecture for distributed knowledge discovery, based on grid computing.

- For business applications, data mining can be integrated into a complex modeling and decision-making process. Reactive business intelligence (RBI) gives an approach, which integrate data mining techniques, modeling techniques, and interactive visualization techniques into a discovery [6].

1.10 Trends in Data Mining

There are several challenging research problems in data mining due to the diversity of data, data mining tasks and data mining approaches. Here is the list of trends in data mining that reflects pursuit of the challenges such as construction of integrated and interactive data mining environments, and design of data mining languages:

1.10.1 Application Exploration

The exploration of data mining for businesses continues to expand for the retail industry by using e-commerce, e-marketing, and e-banking. Upcoming areas of data mining include mobile or wireless data mining and counterterrorism for intrusion detection.

1.10.2 Scalable and Interactive Data Mining Methods

The traditional data mining methods handle very large amount of data efficiently. In contrast to this, recent methods should handle massive amount of data not only efficiently but also interactively. The amount of data being collected continues to grow very rapidly, for that it is essential to have scalable algorithms for individual and integrated data mining. While increasing user interaction, one valuable direction toward improving the overall performance of the mining process is constrain-based mining.

1.10.3 Integration of Data Mining with Database Systems, Data Warehouse Systems, and Web Database Systems

Data mining serves as an essential data analysis component that can be integrated into information processing environment, such as database systems, data warehouse systems, and web. A data mining system should be tightly coupled with database systems and data warehouse systems. In unified framework, one can integrate transaction management, query processing, on-line analytical processing, and on-line analytical mining.

1.10.4 Standardization of Data Mining Query Language

It will smooth the systematic development of data mining solutions, tries to makes improving interoperability among various data mining systems and functions, making available the practice of data mining systems in society, industry, and the relevant field.

1.10.5 Visual Data Mining

It is a novel way to discover knowledge from massive amount of quality data. The systematic study and development of visual data mining techniques will make use of data mining as a tool for analysis of such a huge amount of data.

1.10.6 New Methods for Mining Complex Types of Data

In recent years, most of the problems are solved by using data mining techniques. Still, there are more challenging problems where research frontiers take efforts for solving it. Already many researchers are working on stream data mining, time-series, sequence, graph, spatiotemporal, multimedia, and text data, but still there is very large gap between the needs and technology.

1.10.7 Biological Data Mining

It includes mining DNA and protein sequence, mining high-dimensional microarray data, biological pathway and network analysis, link analysis across heterogeneous biological data, and information integration of biological data by using data mining, which are more interesting topics for biological data mining research.

1.10.8 Data Mining and Software Engineering

The size of software programs is bulky and of different complexity, and modules are integrated from different components developed by different software teams, so the challenging task is to ensure software reliability and robustness of the software. The analysis of such a program is a data mining process. Data mining methodologies are used for automatically detecting the bugs in the software programs, which will be definitely useful in software engineering.

1.10.9 Web Mining

Web mining is the data mining for unstructured data. The information stored on web has different formats. Usually, data on web is not in the structured format, but it is available in unstructured and semi-structured format such as audio data, video data, multimedia data, etc. So processing and analyzing such a large amount of web data is one of the great challenges in front of data miner. Web mining includes mining of the web content, mining of the web log data, and mining of the link.

1.10.10 Distributed Data Mining

Traditional methods are developed to handle the data at a centralized location, which cannot work well for the data distributed at different locations. Advanced distributed data mining methods are expected for handling such homogeneous and/or heterogeneous data available at different locations and in different formats.

1.10.11 Real-Time Data Mining

Many real-life applications involve stream data which requires dynamic data mining models to built in real time. The sophisticated data mining methods are expected to handle for such stream-type data. The examples of stream data are web mining, e-commerce, intrusion detection, stock market analysis, mobile data mining, etc.

1.10.12 Multi-Database Data Mining

The traditional data mining used to discover different patterns in the single database. However, most real-world data and information are spread across multiple databases and multiple tables. Multi-database mining searches for patterns across multiple databases. To handle multiple databases and multiple tables, research is expected for effective and efficient data mining.

1.10.13 Privacy Protection and Information Security in Data Mining

The recorded personal information must be available in digital forms and on the Internet and it poses a threat to our privacy and data security. First to define the privacy by collaborating the experts opinions from different related fields such as technologists, social scientists, law experts and companies and together to work for privacy-preserving data mining [6].

1.11 Classification Techniques in Data Mining

Data classification is the process of organizing data into categories for its most effective and efficient use, for example, predicting tumor cells as benign or malignant, classifying credit card transactions as legitimate or fraudulent, classifying secondary structures of protein as alpha-helix, beta-sheet, or random coil, categorizing news stories as finance, weather, entertainment, sports, etc. A well-planned data classification system makes essential data easy to find and retrieve. Classification is used to predict the class label to specific instance based on the model. Therefore, classification is carried out in two steps: (1) construction of a model and (2) using of a model. In first step, a model is constructed from the sample dataset, where class label is assigned. For assigning class label, some expert person is required, i.e. some form of supervision is required, and hence classification technique is known as supervised learning technique. Output of this step is some form of rules or mathematical or statistical formula. In next step, this generated model is applied on unseen data or test data where class labels are not assigned. In this step, accuracy of the algorithm will be found, and if it is acceptable, then the model is applied on test data for predicting class labels.

1.11.1 Definition of the Classification

Given a Database $D = \{t_1, t_2, \dots t_n\}$ of tuples and a set $C = \{C_1, C_2, \dots C_m\}$, the classification problem is to define a mapping f: $D \rightarrow C$ where each t_i is assigned to one class C_j [7].

1.11.2 Issues Regarding Classification

Here are some of the issues regarding preprocessing the data for classification. The preprocessing steps may be applied on the dataset for better accuracy, efficiency, and the scalability of the classification process.

- **Data cleaning**: The data is processed in order to remove the noise (e.g. by using smoothing techniques) and handle the missing values (missing values are replaced by either most commonly occurring value for that attribute or with the most probable value based on statistics).
- **Relevance analysis**: In the data, many of the attributes may be irrelevant or redundant. Removal of the irrelevant or redundant attributes from the dataset is necessary. The relevance analysis can be used to detect attributes that do not contribute to the classification. The redundancy in the data can be removed by using the technique of correlation analysis. And the irrelevancy of the attributes of the data can be removed by using attribute selection process. The relevance analysis can help improve classification efficiency and scalability.
- **Data transformation and reduction**: For the transformation of the data, normalization may be used. In normalization, we scale all values for a given attribute so that they fall within a small specified range, such as 0.0 to 1.0 or −1.0 to 1.0. The data also transformed by generalizing it to higher level concepts by using concept hierarchy. The other methods can also be used for data reduction, i.e. wavelet transformation, principle component analysis, etc.

1.11.3 Evaluation Methods for Classification

For classification, the following criteria are used to evaluate:

- **Accuracy**: The accuracy of a classifier refers to the ability of a given classifier to correctly predict the class label of new or previously unknown, unseen data. For classifiers, sensitivity, specificity, and precision are useful to measure the accuracy.

- **Speed**: This refers to the computational time required in generating and using the given classifier.
- **Robustness**: This is the ability of the classifier to make correct predictions, when given noisy data or data with missing values.
- **Scalability**: It is concerned with the ability to develop the classifier efficiently, when given large amounts of data.
- **Interpretability**: It is subjective, refers to the level of understanding, and is difficult to assess.

1.11.4 Classifications Techniques

In data mining, lots of classification algorithms were developed. These algorithms are categorized depending on the output of the model. Classification techniques are broadly classified based on the area covered by methods. Following are important categories of classifications:

- Technique Cantered: classification problems may be solved by various methods, such as decision trees, rule-based methods, neural networks, SVM methods, nearest neighbor methods, and probabilistic methods.
- Data-Type Centered: based on the type of data that application created and analyzed, classification algorithms are different. Classification algorithm supports different types of data such as text, multimedia, uncertain data, time series, discrete sequence, and network data.
- Variations on Classification Analysis: there are so many variations of classification techniques available, such as rare class learning, transfer learning, semi-supervised learning, or active learning, and ensemble analysis, which may be used to increase the effectiveness of classification algorithms.

Some important algorithms of classification are as follows.

1.11.4.1 Tree structure

In this type, attributes are arranged in the form of a tree structure. Intermediate nodes in tree are represented by attributes, whereas leaf nodes are represented by class labels. From this tree structure, we may easily generate different classification rules. Some of the important tree structure algorithms are ID3, J48, CART, etc.

1.11.4.2 Rule-based algorithm

In rule-based algorithms, IF-THEN type of production rules are used to represent knowledge. In each rule, conditions are collection of attributes, and statements are represented by class labels. Some of the important rule-based algorithms are Grammar-based genetic programming algorithm (GGP), AprioriC, IF-THEN, genetic algorithm, etc.

1.11.4.3 Distance-based algorithms

This type of algorithm uses similarity or distance between items for classification purpose. Similar items are placed in a particular class. Items in a class are most similar to each other or closest to the centre of that class. One of the well-known examples of this type of algorithm is k nearest neighbors.

1.11.4.4 Neural networks-based algorithms

Neural Network is one of the classification techniques, which is used for information processing system. Neural Network is represented in the form of graph, which consists of a set of nodes (units, neurons, and processing elements) interconnected with each other and having inputs and outputs. Each node has assigned specific weights, which are used to find function computation. Examples of neural network algorithms are multilayer perceptron (with conjugate gradient-based training), a hybrid Genetic Algorithm Neural Network (GANN), etc. [9, 10].

1.11.4.5 Statistical-based algorithms

Based on the quantitative information of characteristics, individual items are divided into groups. These characteristics are inherent in the items from training set of previously labeled items. Some examples of this type of technique are linear discriminate analysis, least mean square quadratic, and kernel.

1.12 Applications of Classifications

Some of potential applications of classification techniques are as follows:

1.12.1 Target Marketing

We know that classification problems are useful for finding future value. This diplomacy is useful for target marketing purpose. In this technique, customer's behaviors are analyzed and future buying perspective is identified.

1.12.2 Disease Diagnosis

Data mining techniques discover patterns from medical records, assign related class label, design model, and use that model for prediction purpose. It is possible to make prediction on the basis of medical information.

1.12.3 Supervised Event Detection

In temporal data, class labels may be associated on the basis of time stamps, which are related to unusual events. In this case, time series classification technique is helpful.

1.12.4 Multimedia Data Analysis

Multimedia data consists of huge amount of complex type of data such as photos, videos, music, etc., on such type of data classification possible. From historical sound track, classification technique identifies future value.

1.12.5 Biological Data Analysis

Biological data generally consist of large volume and small number of elements. Classification techniques are applied on biological data such as DNA or protein sequence and useful to identify properties of sequences.

1.12.6 Document Categorization and Filtering

Many application domains require automatic classifications of documents. Document categorization is one of the use of classification techniques.

1.12.7 Social Network Analysis

There are many uses of classification techniques in social networking technique such as associate labels with underlining nodes. Then, this is used to predict other nodes. Such applications are very useful for predicting useful properties of actors in a social network.

1.13 WEKA: An Effective Tool for Data Mining

Weka is open source software developed at the University of Waikato in New Zealand. "Weka" stands for the Waikato Environment for Knowledge Analysis. The software is freely available at the website [11]. The software is written using object-oriented language Java. Weka can be used to implement various data mining and machine learning algorithms. The algorithms can either be applied directly to a dataset or called from your own Java code. Weka contains different types of modules like data preprocessing, classification, clustering, association rule extraction, and visualization. For our research work, we used Weka version 3.6.7.

1.13.1 Main Features of the Weka

- Data preprocessing tools: In Weka, there are approximately 49 data preprocessing tools available.
- Classification/regression algorithms: Weka provides approximately 76 classification and regression algorithms.
- Clustering algorithms: Weka supports for 8 clustering algorithms.
- Attribute/subset evaluators: In Weka, 15 attribute/subset evaluators are provided and, for feature selection, 10 search algorithms are given.
- Algorithms for finding association rules: Weka also finds association rules by using three algorithms.
- Graphical user interfaces: Weka also provides a graphical user interface through explorer, experimenter, and knowledge flow [12].

1.13.2 Weka Interface

The Weka GUI Chooser provides a starting point for launching Weka's main GUI applications and supporting tools. The GUI Chooser consists of four buttons and four menus (Figure 1.5).

The buttons can be used to start the following applications:

- **Explorer**: An environment for exploring data with WEKA.

It is necessary to open a dataset before starting to explore the data. The explorer contains the following tags:

Preprocess. This tab is used to choose and modify data which are used for processing.

Classify. This tab is useful for training and testing classification techniques or performance regression analysis.

Cluster. This tab is used for clustering analysis.

Associate. In this, tab association analysis is learned.

Select attributes. This tab is used to select the relevant attribute from data.

Visualize. This tab displays data in two dimensional forms.

- **Experimenter**: This button performs experiment and different accompanying statistical tests between learning schemes.

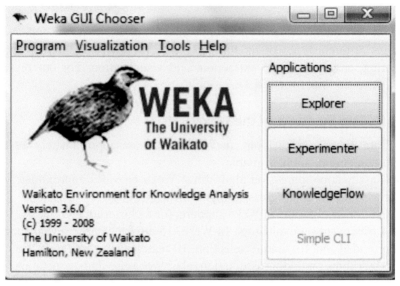

Figure 1.5 WEKA interface.

- **Knowledge Flow**: It is similar to experiment, but it has a drag-and-drop interface and is useful for incremental learning.
- **Simple CLI**: It is a command line interface, which can execute various WEKA commands [12].

1.13.3 Weka for Classification

1.13.3.1 Selecting a classifier

WEKA provides different classification algorithms, which are selected by using the choose button. The selected algorithm is display in text box with its other properties. We may also change the properties of classification algorithm on clicking this box. Each algorithm has its own properties.

1.13.3.2 Test options

The classification algorithm tests result according to test option. There is test option box which contains the following four test modes:

- **Use training set**: The classifier is evaluated on how well it predicts the class of the instances it was trained on.
- **Supplied test set**: Classification algorithm is evaluated on specific test dataset loaded from file. Test dataset is loaded by clicking the Set . . . button and choosing test dataset file.

- **Cross-validation**: Classification algorithm is evaluated by cross-validation fold provided in Folds text field. Default value is 10.
- **Percentage split**: Classification algorithm is evaluated on specific percentage provided in percentage split box. Default value is 66 % field [12].

1.14 What We Aim to Cover Through the Present Book

There are many classification techniques existing, but still there is need to investigate which classification technique gives optimum result when applied to specific domain. As per the survey carried out, there is need for domain specific classification techniques. In the backdrop of web centric applications, we intend to cover the techniques for web mining. The objectives of the present book are as follows:

Objectives of the Book

- To take a review of the existing classification rules.
- To investigate the performance of classification rules for domain-specific data
- To implement classification rules on domain specific data.

Scope and Limitations of the Book

The book aims to report the know-how related to classification of commercial websites with focus on search engine optimization (SEO). Approximately 130 commercial websites data was collected for analysis. As the collected data was not structured, it was preprocessed and filtered to develop the website datasets for analysis. An open source software, Weka, was used for analysis and classification of website datasets.

However, as the websites are dynamic in nature, the classifier works correctly for the datasets collected on specific time period. But it may not give the same results, as the values of the attributes of the website dataset change over the period of time.

Organization of the Book

The book evaluates selected commercial websites from their search engine optimization approach. The organization of the book is as follows:

Chapter 2 presents an in-depth literature review on the classification techniques used in different domains. It is followed by a brief introduction of data mining and its techniques with a special focus on data mining for web. Chapter 3 deals with the preparation of website dataset for analysis. This is followed by data preprocessing and filtering for freezing the parameters for analysis. Chapter 4 presents the techniques used for classification. Accordingly, a comparative analysis of classification techniques is showcased. Chapter 5 gives the summary of this book along with the salient features, conclusion, and applications with future directions.

References

[1] Tan, P.-N. Steinbach, M., and Kumar, V. (2005). *Introduction to Data Mining*. (Upper Saddle River, NJ: Pearson Education).

[2] Linoff, G. S., and Berry, M. J. A. (2011). *Data Mining Techniques for Marketing, Sales and Customer Relationship Management*. 3rd Edn. (New Delhi: Wiley India).

[3] Shawkat Ali, AB. M., and Wasimi, S. A. (2007). *Data Mining: Methods and Techniques*. (Australia: CENGAGE Learning).

[4] (2015). Available: http://cecs.louisville.edu/datamining/PDF/0471228 524.pdf. [Accessed: 26-Mar-2015].

[5] Docs.oracle.com, 'What Is Data Mining?' (2015). [Online]. Available: http://docs.oracle.com/cd/B28359_01/datamine.111/b28129/process.htm #CHDFGCIJ. [Accessed: 13, 14, 29, 30-Mar-2012].

[6] Han, J., and Kamber, M. (2006). *Data Mining: Concepts and Techniques*, 2nd Edn. (Burlington, MA: Morgan Kaufmann Publishers, an imprint of Elsevier).

[7] Dunham, M. H. (2006). *Data Mining Introductory and Advanced Topics*. (Upper Saddle River, NJ: Pearson Education).

[8] Baker, R. S. J. d. (2010). Data mining for education. *Intl. Encyclo. Edu.* 7, 112–118.

[9] Kader, M., and Deb, K. (2012). Neural network-based English alphanumeric character recognition. *Intl. J. Comp. Sci. Eng. Appl. (IJCSEA)* 2.

[10] Prasad, K., Nigam, D. C., Lakhotiya, A., and Umre, D. (2013). Character recognition using Matlab's neural network toolbox. *Intl. J. u- and e-Service, Sci. Technol.* 6.

[11] WEKA at http://www.cs.waikato.ac.nz/~ml/weka

[12] Weka Manual:-http://transact.dl.sourceforge.net/sourceforge/weka/Weka Manual-3.6.0.pdf

2

Current Literature Assessment in Data and Web Mining

Abstract

The literature survey portrayed in the present chapter has displayed the complexity of the data in terms of its size, structure, dimensions, and so on. The literature assessment reveals that the data mining techniques have also evolved with the changes in the data types. Moreover, with the ever-increasing applications on the web-based paradigm, the data mining techniques become too matured to cater to their requirements. Similarity and contrast of mining and classification techniques have also been presented herein. From the multitude of algorithms of data mining, four algorithms are covered and have been focused in the rest of the book.

2.1 Big Data and Its Mining

With the growing number of on-line applications and transactions, there are increasing instances of big data. Data Mining traditionally has been regarded as an important branch of Computer Science and it is gaining paramount significance in the wake of big data. Data mining is very important due to the fact that the raw information is not much useful for processing and needs prefiltering, cleaning, and final processing before it is really useful to be transformed as knowledge. The long journey of the discrete information, facts, figures, and numbers to the useful form of information and knowledge requires all the techniques of the data mining. Since 1970s, there has been quest in the field of data processing, wherein multiple players like software vendors, data base management experts, data analytics, and storage community were involved. However, with the increasing on-line activities, the data has moved from the structured to unstructured form, which has also posed several challenges in front of the technologists working in this niche domain.

In view of the research theme taken up in the present work, which involves data processing along with classification techniques, it is worthwhile to have a look at the basics of the underlying concepts and then proceed to the main algorithms, techniques, and the research work in this area of gaining significance from scientific viewpoint.

2.2 Data-Processing Basics

As per the NIH [1]

> "... 'data' means recorded information, regardless of the form or medium onwhich it may be recorded, and includes writings, films, sound recordings, pictorial reproductions, drawings, designs, or other graphic representations, procedural manuals, forms, diagrams, work flow charts, equipment descriptions, data files, data processing or computer programs (software), statistical records, and other research data."

Today, the speed, complexity, dimensions, and types of data are changing more than ever before. The data processing cycle shown in Figure 2.1 comprises six basic steps.

In the present day, technology has shaped the process of information processing effortless by the aid of electronic information processor. The significance of high-speed data processing and communication to modern society and economy can scarcely be exaggerated. Thomas Friedman, in The World is Flat, argues that they have wrought a more profound revolution change in communication and trade than did the Gutenberg printing press—and have changed the world permanently in far less time [2].

Data mining the core theme of the present research work is an offshoot of the data processing owing to the requirement of intelligent processing with the on shot of the raw data. Extracting the intelligence, patterns, and trends necessitated the mining process; a brief overview of the same is given in the following section.

2.3 Data Mining

There exist various notions of Data Mining. The field is originated due to the fact that the data is now being generated at an awesome rate. The storage capability due to new means such as cloud computing as well as the increased

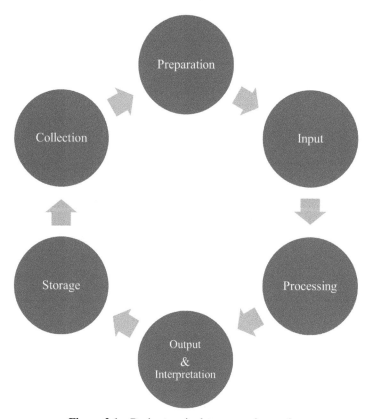

Figure 2.1   Basic steps in data processing cycle.

expectation of the consumers to have processed data has brought in a sea change in the field of data mining. Literature survey reveals the following well-accepted definitions of Data Mining:

"Data mining is the analysis of data for relationships that have not previously been discovered." [2].

"Sifting through very large amounts of data for useful informa-tion." [3].

"Data mining is a powerful new technology with great potential to help companies focus on the most important information in the data they have collected about the behavior of their customers and potential customers. It discovers information within the data that queries and reports can't effectively reveal." [4].

"The extraction of useful, often previously unknown information from large databases or data sets." [5].

"Data mining refers to the systematic software analysis of groups of data in order to uncover previously unknown patterns and relationships." [6].

"A class of database applications that look for hidden patterns in a group of data that can be used to predict future behavior." [7].

"Data mining is the process of accumulating large amounts of data and then combing through it to identify useful information." [8].

"The examination of large amount of information stored in a computer in order to look for patterns, changes, etc." [9].

"Data mining is defined as exploring and analyzing detailed business transactions. It implies 'digging through tons of data' to uncover patterns and relationships contained within the business activity and history." [10].

"Data mining is the process of finding anomalies, patterns and correlations within large data sets to predict outcomes." [11].

"The process of discovering meaningful correlations, patterns and trends by sifting through large amounts of data stored in repositories." [12].

"Data mining also can be defined as the computer-aid process that digs and analyzes enormous sets of data and then extracting the knowledge or information out of it." [13].

"The use of computer software to examine large volumes of data – for example the sales records of a retail business – in order to extract trends and relationships that may be of use in planning the business." [14].

It is interesting to see the origin of the data mining and pioneering work carried out by few research groups.

2.4 Pioneering Work

A brief history of data mining may be explored on web and has been covered extensively in Ref. [15]. The initial emphasis of research work in data mining way back in 1970s to 1990s was on drug approval process for the Food and

Drug Administration and the creation of credit approval curves for credit card companies and banks [16]. Unfortunately, the exact timeframe of conceiving of conceptions of data mining is not obvious. Early 1980s saw its appearance at the instance of overwhelming data generation, storage, processing, and devising patterns out of it [17, 18]. The main essence of data mining is to come out with a model out of the given data which leads to metadata from the mere data. Few similar terms also used to imply Data Mining are: "Exploratory data analysis", "Data driven discovery" and "Deductive learning".

2.5 Algorithms Used in Data Mining

With the evolution of data types, structures and dimensions, the tools, techniques, and means for processing them have also immensely improved. New algorithms are now coming to the forefront for data mining applications.

Basically, a data mining algorithm comprises heuristics and reckoning that generates a model from the data. In order to produce such a model, the algorithm first investigates the data and nurtures explicit types of patterns. The algorithm employs the results of this investigation to define the optimal parameters for creating the mining model. These parameters are then applied across the entire dataset to extract actionable patterns and detailed statistics [19]. The main components of the data mining algorithms are [20]:

- Task
- Structure (functional form) of model or pattern
- Score function to judge quality of fitted model or pattern,
- Search or Optimization method
- Data Management technique storing, indexing, and retrieving data.

Few substantively used data mining algorithms have been summarized in Figure 2.2. As per [21], out of them, the following ten have been found to be used pronouncedly in the field:

- "C4.5
- k-Means
- SVM
- Apriori
- EM
- PageRank
- AdaBoost
- kNN
- Naive Bayes
- CART"

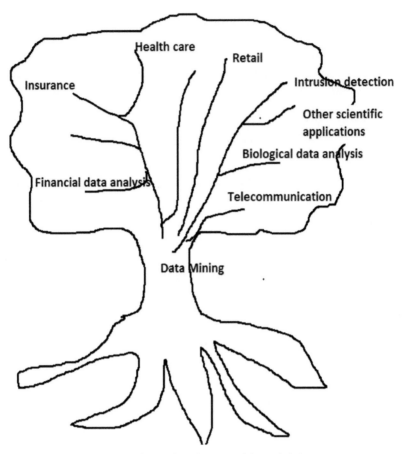

Figure 2.2   Application tree of data mining.

The main reason behind the enhanced interest in data mining algorithms is the increasing ability to track and collect large amounts of data with the use of current hardware technology [22].

With the complexity of data, new techniques such as particle Swarm Optimizers, which are inherently distributed algorithms where the solution for a problem emerges from the interactions between many simple individual agents called particles, are being used [23]. With recent technological advances, shared memory parallel machines have become more scalable, and offer large main memories and high bus bandwidths. They are emerging as good platforms for data warehousing and data mining. In [24], authors have focused on shared memory parallelization of data mining algorithms.

They have developed a series of techniques for parallelization of data mining algorithms, including full replication, full locking, fixed locking, optimized full locking, and cache-sensitive locking.

The following part of the chapter focuses on the research undertaken on the algorithms specifically used in this book. But, prior to that, it would be worthwhile to cover the comparison of classification and mining since the present research work is inseparable amalgamation of the duo.

2.6 Classification and Mining

There are two forms of data analysis that can be used for extracting models describing important classes or to predict future data trends. These two forms are as follows:

- Classification
- Prediction

Classification models predict categorical class labels; and prediction models predict continuous valued functions [25].

Classification is a specialized domain of the data mining. Classification consists of predicting a certain outcome based on a given input. In order to predict the outcome, the algorithm processes a training set containing a set of attributes and the respective outcome, usually called goal or prediction attribute. The algorithm tries to discover relationships between the attributes that would make it possible to predict the outcome [26]. Choosing the proper data mining method is one of the most critical and difficult tasks in the KDD process. In [27], a conceptual map of the most common data mining techniques has been proposed. However, it is concluded that there is not a unique and consensual classification of data mining methods in the literature. However, the algorithms used in the present work have been discussed hereafter.

2.7 Performance Metrics of Classification/Mining

Metrics are some parameters or measures of quantitative assessment used for measurement or comparison in a given context. Data mining metrics may be defined as a set of measurements which can help in determining the efficacy of a Data mining Method/Technique or Algorithm. They are important to help take the right decision as choosing the right data mining technique or algorithm [30]. The following is the list of data mining metrics:

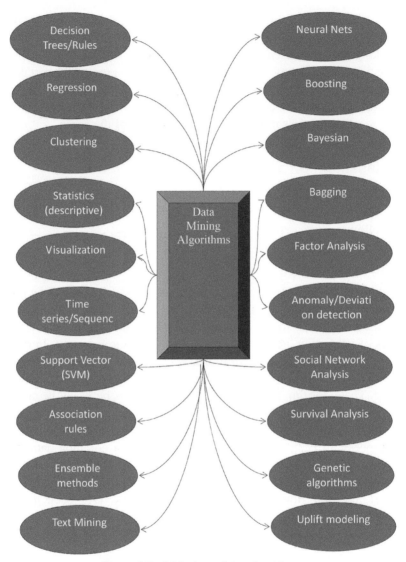

Figure 2.3   Main data mining algorithms.

- Usefulness
- Return on investment
- Accuracy
- Space/Time [31].

The above data mining metrics are shown in Figure 2.4.

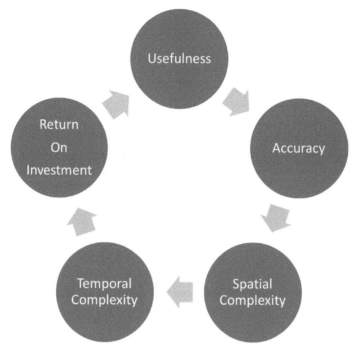

Figure 2.4   Metrics of data mining algorithm.

2.8 Data Mining for Web

The theme of the present research work, i.e. **Web Data Mining**, is a technique used to crawl through various web resources to collect required information, which enables an individual or a company to promote business, understanding marketing dynamics, new promotions floating on the Internet, etc. There is a growing trend among companies, organizations, and individuals alike to gather information through web data mining to utilize that information in their best interest [31].

2.9 Categories of Web Data Mining

The broad categories of the web data mining include web content mining, web structure mining, and web usage mining. Web usage mining refers to the discovery of user access patterns from Web usage logs. Web structure mining tries to discover useful knowledge from the structure of hyperlinks. Web content mining aims to extract/mine useful information or knowledge from web page contents. This tutorial focuses on Web Content Mining [22]. It is summarized in Figure 2.5.

Web mining is defined as the process of extracting useful information from World Wide Web data. Two different approaches are taken to define the web mining: (1) "Process-centric view", which defines web mining as a sequence of task. (2) "Data-centric view", which defines web mining in terms of the types of web data that is being used in the mining process. Mainly web mining is divided into three types: (1) Web Content Mining, (2) Web usage mining, and (3) Web Structure Mining. Web content mining is the process of extracting the useful information from the web document. To perform this task, there are two main approaches like agent-based approach and database approach.

Web usage mining is the process of mining the important information from web history. Web usage mining process can be divided into three stages: (1) Pre-processing, (2) Pattern discovery, and (3) Pattern analysis. In preprocessing stage, data is cleaned and partitioned into a set of user's transaction that represents the activity of each user during the visiting of different sites. In Pattern discovery-stage database, machine learning and statistical operations are performed to obtain the hidden patterns that reflect the behavior of user. Pattern analysis: In this, discovered patterns are further processed, filtered, and analyzed in user model that can be used as input to applications such as visualization tools and report generation tools.

Web structure mining is a very difficult task. It is the process of analyzing the hyperlink and extracting important information from it. It is also used to mine the structure of document, and analyze the structure of page to describe the HTML or XML usage. The goal of the Web Structure Mining is to generate the structural summary about the Web site and Web page. Web Structure mining will categorize the Web pages and generate the information like the similarity and relationship between different Web sites. This type of

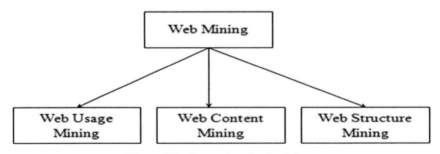

Figure 2.5   Categories of web mining.

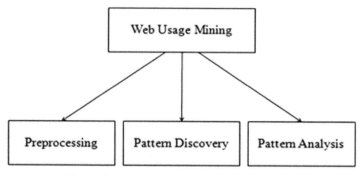

Figure 2.6 Sequence of events of web mining.

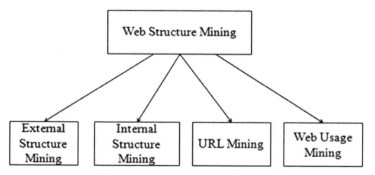

Figure 2.7 Web structure mining.

mining can be performed either at document level that is referred to as intra-page or at hyperlink level that is referred to as inter-page mining. Owing to the importance of this mining technique, there have been several algorithms proposed to solve this. In this paper, we will describe and analyze web structure mining algorithms and identify their strengths and limitations [23].

The literature survey reveals many papers [33–39] from different perspectives of web mining. The underlying techniques play an important role in web mining and the same is discussed in the following section. The techniques discussed below are essentially pertaining to one used in the implementation of the current book.

2.10 Radial Basis Function Networks

The idea of Radial Basis Function (RBF) Networks derives from the theory of function approximation. "Their main features are [28]:

- They are two-layer feed-forward networks.
- The hidden nodes implement a set of radial basis functions (e.g. Gaussian functions).
- The output nodes implement linear summation functions as in an MLP.
- The network training is divided into two stages: first the weights from the input to hidden layer are determined, and then the weights from the hidden to output layer are determined.
- The training/learning is very fast.
- The networks are very good at interpolation."

Popularized by Moody and Darken (1989), RBF networks have proven to be a useful neural network architecture. The major difference between RBF networks and back propagation networks (that is, multi-layer perceptron trained by Back Propagation algorithm) is the behavior of the single hidden layer [29]. Other applications of RBF are discussed in depth in [40–42].

2.11 J48 Decision Tree

Decision tree J48 is the implementation of algorithm ID3 (Iterative Dichotomiser 3) developed by the WEKA project team [43]. Invented by J. Ross Quinlan, ID3 employs a top-down greedy search through the space of possible decision trees. The attribute that is most useful for classifying examples (attribute that has the highest Information Gain) has been selected. More details regarding the same are covered in [44–47].

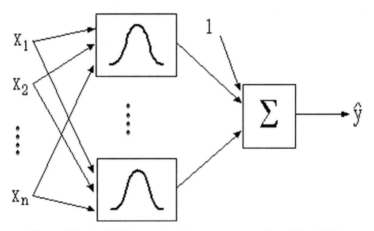

Figure 2.8   An RBF network with one output retrieved from [29].

2.12 Naive Bayes

Despite its simplicity, the Naive Bayes learning scheme performs well on most classification tasks, and is often significantly more accurate than more sophisticated methods [48]. The Naive Bayes classifier is a generative model wherein the oversimplifying assumption is made that the inputs are independent, given the label [49]. Details as regards to application of Naive Bayes Regression for web mining are covered widely in the literature [50–54].

2.13 Support Vector Machine (SVM)

SVM was introduced in COLT-92 by Boser, Guyon and Vapnik. Theoretically well-motivated algorithm, developed from Statistical Learning Theory (Vapnik and Chervonenkis) since the 60s, became so popular that many classifiers have been built with the same [55]. SVM applications in context with the web data mining are discussed in [56–59].

2.14 Conclusion and Way Forward

The literature survey portrayed in the present chapter has displayed the complexity of the data in terms of its size, structure, dimensions, and so on. The data mining techniques have also changed with the changes in the data types. Moreover, with the ever-increasing applications on the web-based paradigm, the data mining techniques also mature to cater to their requirements. Similarity and contrast of mining and classification techniques have also been presented herein. From the multitude of algorithms of data mining, four algorithms covered under Figures 2.6–2.8 have been focused in the further part of the book.

References

[1] 2015. [Online]. Available at: http://www.aplu.org/members/councils/re search/cor-archive/cor-2013/2013-the-changing-nature-uses-of-data.pdf. [accessed Apr 27, 2015].
[2] 2015. [Online]. Available at: https://sfog.lbl.gov/sites/all/files/lbnl-63339.pdf. [accessed Apr 27, 2015].
[3] Searchsqlserver.techtarget.com, "What is data mining? Definition from WhatIs.com", 2015. [Online]. Available at: http://searchsqlserver.tech target.com/definition/data-mining. [accessed Feb 04, 2015].

[4] BusinessDictionary.com, "What is data mining? definition and meaning", 2015. [Online]. Available at: http://www.businessdictionary.com/definition/data-mining.html. [accessed Feb 04, 2015].

[5] Laits.utexas.edu, "Data Mining", 2015. [Online]. Available at: http://www.laits.utexas.edu/~anorman/BUS.FOR/course.mat/Alex/. [accessed Feb 04, 2015].

[6] TheFreeDictionary.com, "Data mining", 2015. [Online]. Available at: http://www.thefreedictionary.com/data+mining. [accessed Feb 04, 2015].

[7] Investinganswers.com, "Data Mining Definition & Example | Investing Answers", 2015. [Online]. Available at: http://www.investinganswers.com/financial-dictionary/businesses-corporations/data-mining-1563. [accessed Feb 04, 2015].

[8] Webopedia.com, "What is Data Mining? Webopedia", 2015. [Online]. Available at: http://www.webopedia.com/TERM/D/data_mining.html. [accessed Feb 04, 2015].

[9] TrackMaven, "Data Mining Definition—at TrackMaven.com", 2015. [Online]. Available at: http://trackmaven.com/marketing-dictionary/data-mining/. [accessed Feb 04, 2015].

[10] Dictionary.cambridge.org, "Data mining—definition in the Business English Dictionary—Cambridge Dictionaries Online", 2015. [Online]. Available at: http://dictionary.cambridge.org/dictionary/business-english/data-mining. [accessed Feb 04, 2015].

[11] Pcmag.com, "data mining Definition from PC Magazine Encyclopedia", 2015. [Online]. Available at: http://www.pcmag.com/encyclopedia/term/40813/data-mining. [accessed Feb 04, 2015].

[12] Sas.com, "What is Data Mining?", 2015. [Online]. Available at: http://www.sas.com/en_us/insights/analytics/data-mining.html. [accessed Feb 04, 2015].

[13] Modeling, D. (2012) [Online]. "Data Mining—Big Data Analytics—Gartner", *Gartner IT Glossary*. Available at: http://www.gartner.com/it-glossary/data-mining. [accessed Feb 04, 2015].

[14] ZenTut, "What is Data Mining", 2015. [Online]. Available at: http://www.zentut.com/data-mining/what-is-data-mining/. [accessed Feb 04, 2015].

[15] Sewell, M. 2015. [Online] "Data Mining", Datamining.martinsewell.com. Available at: http://datamining.martinsewell.com/. [accessed Feb 04, 2015].

[16] Researcher.watson.ibm.com, "Knowledge Discovery and Data Mining—IBM", 2015. [Online]. Available at: http://researcher.watson.ibm.com/researcher/view_group.php?id=144. [accessed Feb 04, 2015].

[17] Anderson.ucla.edu, "Data Mining: What is Data Mining?", 2015. [Online]. Available at: http://www.anderson.ucla.edu/faculty/jason.frand/ teacher/technologies/palace/datamining.htm. [accessed Feb 04, 2015].

[18] Furnas, A. (2012). [Online]. "Everything you wanted to know about data mining but were afraid to ask", *The Atlantic*, Available at: http://www.theatlantic.com/technology/archive/2012/04/everything-you-wanted-to-know-about-data-mining-but-were-afraid-to-ask/25538 8/. [accessed Feb 04, 2015].

[19] Msdn.microsoft.com, "Data Mining Algorithms (Analysis Services—Data Mining)", 2015. [Online]. Available at: https://msdn.microsoft.com/ en-us/ms175595.aspx. [accessed May 02, 2015].

[20] 2015. [Online]. Available at: http://www.cedar.buffalo.edu/~srihari/CSE 626/Lecture-Slides/Ch5-Part1-SystematicOverview.pdf. [accessed May 02, 2015].

[21] Big Data Analytics, Data Visualization and Infographics, "Top most algorithms used in Data Mining", 2012. [Online]. Available at: https: //cloudcelebrity.wordpress.com/2012/05/04/top-most-algorithms-used-in-data-mining/. [accessed May 02, 2015].

[22] Bertino, E., Fovino, I., and Provenza, L. (2005). "A framework for evaluating privacy preserving data mining algorithms*", *Data Mining Knowledge Discov.* 11 (2), 121–154.

[23] Jin, R., Yang, G., and Agrawal, G. (2005). "Shared memory parallelization of data mining algorithms: techniques, programming interface, and performance", *IEEE Trans. Knowledge Data Eng.* 17 (1), 71–89.

[24] 2015. [Online]. Available at: http://courses.cs.washington.edu/courses/ csep521/07wi/prj/leonardo_fabricio.pdf. [accessed May 02, 2015].

[25] Tutorialspoint.com, "Data Mining Classification & Prediction", 2015. [Online]. Available at: http://www.tutorialspoint.com/data_mining/dm_ classification_prediction.htm. [accessed May 02, 2015].

[26] 2015. [Online]. Available at: http://www.iemss.org/iemss2010/papers/S 23/S.23.03.Choosing%20the%20Right%20Data%20Mining%20Techn ique%20-%20Classification%20of%20Methods%20and%20Intelligent %20Recommendation%20-%20MIQUEL%20SANCHEZ-MARRE. pdf.[accessed May 02, 2015].

[27] 2015. [Online]. Available at: http://www.cs.bham.ac.uk/~jxb/NN/l12.pdf. [accessed May 02, 2015].

[28] Reference.wolfram.com, "Radial Basis Function Networks", 2015. [Online]. Available at: http://reference.wolfram.com/applications/neural networks/NeuralNetworkTheory/2.5.2.html. [accessed May 02, 2015].

[29] Arteaga, C., and Marrero, I. (2013). "Universal approximation by radial basis function networks of Delsarte translates", *Neural Networks* 46, 299–305.

[30] 2015. [Online]. Available at: http://himadri.cmsdu.org/documents/data mining_metrics.pdf. [accessed Feb 05, 2015].

[31] Slideshare.net, "Ch 1 Intro to Data Mining", 2015. [Online]. Available at: http://www.slideshare.net/sushil.kulkarni/ch-1-intro-to-data-mining-presentation. [accessed Feb 05, 2015].

[32] Web-datamining.net, "Web Data Mining—An Introduction", 2015. [Online]. Available at: http://www.web-datamining.net/. [accessed Feb 05, 2015].

[33] Stoffel, K. (2009). "Web + Data Mining = Web Mining", *HMD Praxis der Wirtschaftsinformatik*, 46 (4), 6–20.

[34] Spiliopoulou, M. (2000). "Web usage mining for Web site evaluation", *Commun. ACM*. 43 (8), 127–134.

[35] Sivakumar, P. (2015). "Effectual web content mining using noise removal from web pages", *Wireless Pers Commun*, 2015.

[36] Xu, G., Yu, J., and Lee, W. (2013). "Guest editorial: social networks and social web mining", *World Wide Web*, 16 (5–6), 541–544.

[37] Jiang, D., Leung, K., and Ng, W. (2015). "Query intent mining with multiple dimensions of web search data", *World Wide Web*.

[38] Liu, Y. (2014). "Fuzzy-Clustering Web based on Mining", *JMM*, 9 (1).

[39] Wang, T., and Fan, G. (2013). "The development and design of intelligent web site based on web usage mining", *AMR*, 718–720, 2074–2079.

[40] Xiao, Y., Liu, J., Wang, S., Hu, Y., and Xiao, J. (2014). "Multiple dimensioned mining of financial fluctuation through radial basis function networks", *Neural Comput. Applic*, 26 (2), 363–371.

[41] Billings, S., Wei, H., and Balikhin, M. (2007). "Generalized multiscale radial basis function networks", *Neural Networks*, 20 (10), 1081–1094.

[42] Wilkins, M., Morris, C., and Boddy, L. (1994). "A comparison of radial basis function and backpropagation neural networks for identification of marine phytoplankton from multivariate flow cytometry data", *Bioinformatics* 10 (3), 285–294.

[43] Data-mining.business-intelligence.uoc.edu, "J48 decision tree—Mining at UOC", 2015. [Online]. Available at: http://data-mining.business-intelligence.uoc.edu/home/j48-decision-tree. [accessed May 05, 2015].

[44] Vasudevan, P. (2014). "Iterative dichotomiser-3 algorithm in data mining applied to diabetes database", *J. Comp. Sci.* 10 (7), 1151–1155.

[45] Zeng, H. (2014). "Iterative and non-iterative solution of planar resection", *AMR* 919-921, 1295–1298.

[46] 2015. [Online]. Available at: http://www.cs.sjsu.edu/faculty/lee/cs157b/ .../Chris%20Archibald%20ID3.ppt. [accessed May 05, 2015].

[47] 2015. [Online]. Available at: https://www.rivier.edu/journal/ROAJ-Fall-2012/J674-Slocum-ID3-Algorithm.pdf. [accessed May 05, 2015].

[48] Hoare, Z. (2007). "Landscapes of NaiveBayes classifiers", *Pattern Anal. Appl.* 11 (1), 59–72.

[49] 2015. [Online]. Available at: http://melodi.ee.washington.edu/~halloj3/ classification.pdf. [accessed May 05, 2015].

[50] Calders, T., and Verwer, S. (2010). "Three NaiveBayes approaches for discrimination-free classification", *Data Mining Knowl. Discovery*, 21 (2), 277–292.

[51] Shaw, D., and Chellappa, R. (2013). "Regression on manifolds using data-dependent regularization with applications in computer vision", *Statistical Analy Data Mining*, 6 (6), 519–528.

[52] Dhakar, M., and Tiwari, A. (2014). "Tree-augmented NaiveBayes-based model for intrusion detection system", *Intl. J. Know. Eng. Data Mining*, 3 (1), 20.

[53] Fraley C., and Hesterberg, T. (2009). "Least angle regression and LASSO for large datasets", *Statistical Analysis Data Mining*, 1 (4), 251–259.

[54] Melnykov, V. (2012). "Efficient estimation in model-based clustering of Gaussian regression time series", *Statistical Analy Data Mining*, 5 (2), 95–99.

[55] 2015. [Online]. Available at: http://www.cs.columbia.edu/~kathy/cs4701/ documents/jason_svm_tutorial.pdf. [accessed May 05, 2015].

[56] Lo, S. (2008). "Web service quality control based on text mining using support vector machine", *Expert Syst. Appl.* 34 (1), 603–610.

[57] Das, S., and Saha, S. (2013). "Data mining and soft computing using support vector machine: a survey", *Intl. J. Comput. Appl.* 77 (14), 40–47.

[58] Manimala, K., Selvi, K., and Ahila, R. (2012). "Optimization techniques for improving power quality data mining using wavelet packet based support vector machine", *Neurocomputing* 77 (1), 36–47.

[59] Zhang, W., Pandurangi, A., Peace, K., Zhang, Y., and Zhao, Z. (2011). "MentalSquares: A generic bipolar Support Vector Machine for psychiatric disorder classification, diagnostic analysis and neurobiological data mining", *Intl. J. Data Mining and Bioinformat.* 5 (5), 532.

3

DataSet Creation for Web Mining

Abstract

This chapter has laid emphasis on the paradigm shift of Data Mining to Web Mining owing to the demands from the real world for live data analysis in order to devise valuable, hidden, potential patterns from the large amount of data. The dataset derived in the last chapter demonstrates the data distributed at different locations and in fact is supposedly opening step toward web mining. As presented earlier, web mining implies a methodology to extract the valuable, important pattern or knowledge from the data which is distributed at remote locations or servers. A free shareware tool such as site-analyzer helps in building such a dataset for the purpose of web mining. As demonstrated, the data of approximately 150 commercial websites has been collected and used for the analysis. The site-analyzer tool empowers in gathering useful metrics such as global score, web accessibility, design, texts, multimedia, and networking. This collected data is then further stored in tabular form parameterwise for the purpose of further process of web mining. This is followed by application of preprocessing techniques in order to remove the unexpected and non-relevant data items. As presented in this chapter, sometimes there are issues like not display/hidden data items in case of few commercial websites, which ultimately needs to be removed from the dataset. The values of the parameters are displayed in the form of either 'yes' or 'no'. For such a case, 'yes' is replaced with 1 and 'no' is replaced with 0. Similarly, other parameters are also handled. It is made sure that the data of all these parameters are free from error and preprocessed completely. As further demonstrated in this chapter, filtering techniques are necessitated to remove unwanted columns from the data of all parameters as the same does not play an important role in analysis. All these steps ensure data in standard form and available for further processing and analysis using Weka. Methodology reported in this chapter is a unique one, which can be applied for any potential domain wherein data is emanating from multiple remote sources.

55

3.1 Introduction

A dataset is a collection of data objects, which are described by a number of attributes. These attributes represent the basic characteristics of an object [1]. Once the data is generated, one cannot understand and cannot get any type of information just by looking into the data. We have rich data but information is poor. To get information, it is necessary to process such data. The data mining algorithms use this data for analysis and try to extract the non-trivial, valuable and important hidden patterns and knowledge in the datasets. Actually, the real-world data is dirty, rough, and raw, i.e. it contains errors. When we process the raw data then outcome may lead to wrong results. To convert such raw data into quality data, i.e. free of errors, it is essential to apply the data preprocessing techniques on it such as data cleaning, data integration, data selection, and data transformation. After preprocessing the data, it is free from errors. The preprocessed data can be processed in order to get the quality result, which will help for proper decisions based on its quality output [1, 2].

3.2 Web Mining—Emerging Model of Business

3.2.1 Introduction to Web Mining

Throughout the world, there are about two billion pages created by millions of web page authors and organizations. WWW has become tremendously rich in knowledge base. Not only the knowledge comes from the content of the web pages themselves, but also the unique characteristics of the web, like its hyperlink structure and its diversity of content and languages. After analysis of these characteristics of the web pages, one can get interesting patterns and also gain new knowledge. Such created new knowledge can be used for the improvement of user's efficiency and effectiveness in searching details on the web. At the same time, it can be used for decision-making or business management.

The content of the web is unstructured and dynamic, its size is also dynamic and its nature is multilingual. Extraction of the useful information from such heterogeneous data is a challenging research problem. In addition to this, web generates a huge amount of data in formats that contain valuable information. For example, Web server logs' information about user access patterns can be used for improving Web page design or information personalization.

Machine learning techniques have been used by researchers to solve the problem of discovering the patterns and knowledge from such a large amount of heterogeneous data. In business and scientific domain, the techniques of artificial intelligence and machine learning are applied and data mining research plays an important role in this area. Similarly, in information retrieval and text mining applications, machine learning techniques are also used. The various activities and efforts in this area are referred to as Web mining. The term Web mining was coined by Etzioni (1996) to denote the use of data mining techniques to automatically discover Web documents and services, extract information from Web resources, and uncover general patterns on the Web. Traditionally, Web mining research has been extended to cover the use of data mining and similar techniques to discover resources, patterns, and knowledge from the Web and Web-related data such as Web usage data. Hsinchun Chen and Michael Chau defined Web mining as "the discovery and analysis of useful information from the World Wide Web". The term Web Mining refers to computational processes that aim to discover useful information or knowledge from data on the Web. Based on the data used in the mining process, Web Mining can be subdivided into three non-disjoint types: (i) Web structure mining, (ii) Web content mining, and (iii) Web usage mining. Web structure mining discovers knowledge from the hyperlink structure of the Web. Web content mining extracts useful information from page contents and Web usage mining extracts knowledge from the usage patterns that people leave behind as they interact with the Web [3].

Before the introduction of the Internet concept, market was totally dependent on the traditional methods. In traditional methods, customer approaches the shop and purchases items for their daily needs. To increase profit margin, business people attract more and more customers by launching different innovative schemes and declare the discounts on purchasing items. To do this, the business people contact customers through the medium of advertisement.

However, as the Internet concept is launched, business people try to reach the customers through the computer. These people develop their own websites and make everything available to the customer. The customer just visit a particular website for details. This concept is extended and now people make any type of transactions through Internet by using wired connection or wireless connection. Recently, the trend has changed from traditional to online purchasing of any item. The online transactions are possible due to e-commerce and e-banking. To increase profit and attract more customers, business people try to reach their customers through web.

World Wide Web facilitates connection of any customer anytime and anywhere. The websites developed by business people are growing day by day. All business people do their business through the web. But the question is that which customers are happy with which websites? Why one website is popular? What parameters have effect on attracting more customers? Which business people provide quick response? So many questions arise for the smooth running of the business.

Many parameters are playing important roles for the assessment of the business houses. Websites are assessed with respect to web page loading speed, readability, broken-link, navigation, proper URL, graphics, etc. Similarly, the parameters related to website design are also considered for the present work. For present work, "website design-analysis tool" is used. Which is available on website http://www.site-analyzer.com. In website design analysis, we recorded the values of the parameters like accessibility, design, texts, multimedia, and networking. Website data of approximately 125–150 social networks is recorded. The web accessibility has sub-parameters such as page weight, compression, ease caching and download time. For these parameters, actual values are also recoded. Regarding URL, we collected the data of the parameters like URL rewriting, domain length, domain age and whois (creation date, expiration date, and last update). Networking parameter in turn has sub-parameters such as number of links, link juice, single links, links without underscores, title attribute, linking, and reliable link. The details of these networking parameters are given as follows:

- **Link juice**: It is the ratio between internal and external links. It helps to recognize whether the site has gateway or not, if yes it has to redirect to different sites. It is recommended that the ratio must be greater than 50% [4].
- **Single links**: Website having more number of duplicate links is not good for SEO. It is obtained by taking the ratio of single links and duplicate links [4].
- **Links without underscore**: It is recommended not to use underscore in a link. Instead of using underscore, one should use hyphen because search engine did not recognize it as word separator [4].
- **Title attributes**: The title attribute allows one to correlate keywords to the contents of the particular page [4].
- **Linking**: This attribute indicates to search engine whether or not this link should be followed. If there is too many "no follow" link juice, then it reduces the importance of the other links. It is recommended to avoid defining more than five "no-follow" links on the same page [4].
- **Reliable links**: It is related to the number of reliable links. That is to say, the number of links which search engine can crawl [4].

3.3 Tools Used for Acquisition of Parameters

In present work, the site-analyzer tool is used to collect the information about various types of parameters of each social network websites. Site-Analyzer performs a multi-criteria SEO analysis. Site-Analyzer is a Search Engine Optimization (SEO) tool to analyze the website. One quote that is displayed on home page of site-analyzer tool is **'Website optimization is not only an SEO improvement, it's making the web better'**. This tool is a complete website analysis tool used to improve SEO. Site-Analyzer is an optimization tool for the websites. This SEO tool will allow analyzing website and generating a multi-point audit sorted by category such as accessibility, design, texts, multimedia, and networking.

The analysis report is made of more than 50 criteria based on the optimization of several factors such as server configuration, HTML tagging, text content, multimedia content, internal and external networking, and page popularity.

This SEO analysis allows us to check the performance of the website and helps to improve the visibility of the website for search engines. In this way, it helps for improving the natural ranking of the site and also increases the potential number of visitors to website. All these criteria are a powerful asset for the website.

- **SEO Analysis**: As per the standards of search engines, the pages are analyzed and keyword density, links, meta tags, and other related elements, which influence natural SEO, are optimized [4].
- **Performance**: By analyzing the server configuration, multimedia content, and the loading speed of the HTML page, it is possible to improve the performance of various websites [4].
- **Design**: In order to improve the user's experience, it is recommended to control the design of the page through in-depth analysis of HTML code, the general web design, style sheets, and Javascript [4].
- **Compatibility**: Website should be optimized in order to make it compatible with almost all devices such as desktops, laptops, ts, and smart phones. Whatever the configuration of websites, it should be accessible to the users throughout the world [4].

Figure 3.1 shows the home screen of the site-analyzer tool.

One text box is provided with button 'ANALYZE' waiting for web address of the website. Once the full website address is provided after pressing the ANALYZE button, the site-analyzer starts analysis of the website. After some time, it displays the results.

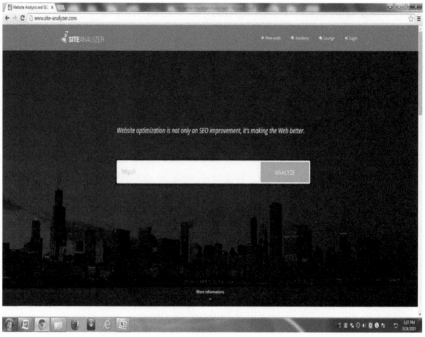

Figure 3.1   Home page of site-analyzer website.

The web analyzer displays the analysis report, which includes overview, accessibility, design, texts, multimedia, and networking. All the values of these parameters are displayed in percentage.

- **Overview**: The overview parameter just displays the global score in percentage. Figure 3.2 displays the global score of website obtained by using the site-analyzer tool [4].
- **Accessibility**: It is one of the important aspects in a website's SEO strategy. Anyone can visualize the accessibility from different angles. From a human point of view, it is very important to be able to analyze if the content on the website is readable by a human eye. And the same person will be able to access all the contents without any problem. From a machine's point of view, the technical accessibility to a website is important to be able to read and crawl its content [4].
- **Design**: Web page design is a technical aspect. It is defined by the structure of its source code. The design provides the content of the page with a certain structure, which allows the information to be easily readable by search engines. The elements of the header are considered

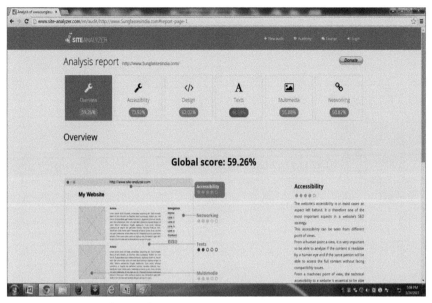

Figure 3.2   Overview displays the global score.

for the analysis of the design of the website, which is invisible to the user but necessary for the proper execution of the web page [4].

- **Texts**: Internet is a wealth of information because it gave access of all the information on web to the user at anytime, anywhere. To attract more and more visitors, it is very essential to provide them quality textual content that will interest them. The quality of the textual content is totally based on the analysis of keyword density, the semantics of your page, and the formatting of the text. The word field used in the web page is given by this analysis [4].
- **Multimedia**: This analysis gives the specification of the different types of media elements used in the web page. When non-textual contents are available on web page, it affects website's loading speed [4].
- **Networking**: A page's networking contains the details such as internal and external links, indexing statistics and technical details for search engines. For healthy navigation, it is most important to have a good internal networking. It is necessary to send search engines information concerning the website, so they can crawl and map it for better visibility [4].

Table 3.1 shows sample of actual collected overall data.

Table 3.1   Actual collected overall data

Sr. No.	Name of the URL	Global Score %	Accessibility %	Design %	Texts %	Multimedia %	Networking %	Date	Time
Fashion and Lifestyle									
1	www.Sunglassesindia.com	59.23	73.72	62.02	46.67	55.88	50.82	13-03-2012	3.02 pm
2	www.Brandsndeals.com	56.01	71.29	52.73	52.15	38.23	58.19	13-03-2012	3.10 pm
3	www.Majorbrands.in	55.91	56.39	66.67	52.03	55.88	42.62	13-03-2012	4.14 pm
4	www.Pantaloons.com	47.9	46.42	87.43	29.69	35.88	14.75	13-03-2012	5.05 pm
5	www.Shoppersstop.com	51.31	64.56	65.46	57.13	24.12	27.87	13-03-2012	5.11 pm
6	www.Inkfruit.com								
7	www.UtsavSarees.com								
8	www.Globus.com	59.61	60.81	54.64	66.36	74.71	49.18	13-03-2012	5.17 pm
9	www.zovi.com								
10	www.BagsKart.com								
11	www.Fetise.com	44.62	49.46	59.29	45.4	17.06	35.25	13-03-2012	5.24 pm
12	www.Orosilber.com	53.47	57.91	62.39	38.77	24.12	68.03	13-03-2012	5.28 pm
13	www.Violetbag.com	61.23	74.63	64.25	48.51	61.18	51.64	13-03-2012	5.34 pm
14	www.Bags109.com								
15	www.Myntra.com	55.11	53.11	70.81	29.24	61.18	52.46	13-03-2012	5.38 pm
16	www.Elitify.com	57.22	64.87	75.96	46.28	24.12	52.46	14-03-2012	8.40 am
Custom designed T-shirt, mug, calendar, etc.									
17	www.Zoomin.com								
18	www.Snapfish.in								
19	www.Picsquare.com	51.66	60.71	52.73	61.08	31.18	45.08	14-03-2012	8.50 am
Footwear									
20	www.Yebhi.com	56.68	48.95	76.5	41.26	63.53	45.08	14-03-2012	8.55 am
21	www.MetroShoes.com								
22	www.BataShoes.com								
23	www.BeStylish.com	61.81	64.87	72.13	38.46	55.88	67.21	14-03-2012	9.00 am

3.3.1 Accessibility

Figure 3.3 shows the details of web accessibility parameter obtained by using the site-analyzer tool [4].

The accessibility consists of following parameters:

- **General data**:
  - **Page weight**: It is recommended not to go over 50ko for the pages (without considering images and other media. This only concerns the weight of the HTML code). This criterion is taken into consideration by the search engines in their ranking algorithms [4].
  - **Compression gzip/deflate**: Compression of the web pages allows to optimize web page loading time by reducing the amount of data to be downloaded. It is activated on server-side [4].
  - **Page caching**: Caching pages will reduce their loading time by storing them for time being on the user's device. The caching delay has to be set depending on the page updating frequency [4].
  - **Download time**: This criterion allows to see the average time needed by visitors to download the full content of the web page such as images, javascripts, and CSS files based on the page

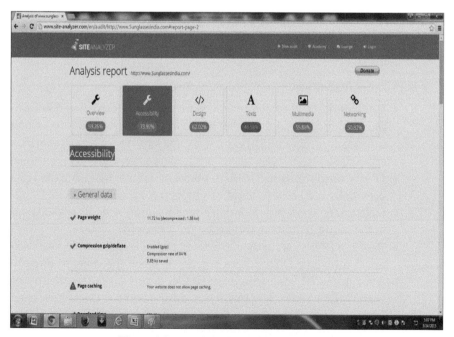

Figure 3.3 Accessibility with its parameters.

weight and global average connection speed. The main aim being to reduce the time needed to display the information on visitor's screens [4]. The Table 3.2 shows sample of actual collected web accessibility data.

- **URL Optimization**:
  - **www redirection**: It is necessary to redirect the web pages to http://www or http:// (with or without www) to avoid duplicate content. The website should be easy to get through a single address [4].
  - **URL Rewriting**: URL rewriting allows you to rewrite this kind of link: http://www.car.com/page.php?id=2&category=3 to a more user-friendly one like: http://www.car.com/sedan-diesel-2009. Also this option allows to add keywords to explain the concerned page [4].
  - **Domain length**: A short domain name anyone can remember easily. If the domain name is long, it has less chance to be quickly memorized by the user [4].
  - **Domain age**: Anyone can see when the domain was created. It will be more reliable when it is older [4].
  - **Whois**: The WHOIS database informs about the owner, the creation date and the expiration date of a domain. It also gives other miscellaneous information like the owner's contact information, the registrar, etc. [4].

- **Server configuration**:
  - **Environment**: It is recommended to hide the information related to the technologies used on web server, to avoid giving out too much information that could be used by hackers to gain unauthorized access [4].
  - **IPv6 Compatibility**: The IPv6 protocol will soon replace the IPv4 protocol. Web server should be prepped for the same [4].
  - **DNS zone**: The DNS (Domain Name System) zone routes the user to the requested active service on the server [4].
  - **Reverse DNS**: Part of the DNS zone, the PTR (pointer) record used to set a reverse DNS allows servers to reach the domain directly through the IP. This record is used to improve the traceability of server's IP address. It is mainly used for email servers, which need to take into consideration a trust factor to improve the delivery rate of the emails and avoid being flagged as spam [4].

Table 3.2 Actual collected web accessibility data

Sr. No.	Name of the URL	Page Weight	Compression %	Page Caching	Download Time	Date	Time
Fashion and Lifestyle							
1	www.Sunglassesindia.com	12.32	87	Not allowed	684.17	13/03/2012	3.02 pm
2	www.Brandsndeals.com	11.59	65	Not allowed	1.26	13/03/2012	3.10 pm
3	www.Majorbrands.in	19.11	89	Not allowed	2.83	13/03/2012	4.12 pm
4	www.Pantaloons.com	11.69	80	Not allowed	6.3	13/03/2012	5.05 pm
5	www.Shoppersstop.com	30.97	87	Not allowed	2.4	13/03/2012	5.11 pm
6	www.Inkfruit.com						
7	www.UtsavSarees.com						
8	www.Globus.com	2.34	64	Not allowed	2.34	13/03/2012	5.17 pm
9	www.Zovi.com						
10	www.BagsKart.com						
11	www.Fetise.com	11.43	74	Not allowed	1.99	13/03/2012	5.24 pm
12	www.Orosilber.com	17.87	72	Not allowed	3.37	13/03/2012	5.28 pm
13	www.Violetbag.com	12.33	85	Not allowed	810.93	13/03/2012	5.34 pm
14	www.Bags109.com						
15	www.Myntra.com	136.64	51	Not allowed	1.77	13/03/2012	5.38 pm
16	www.Elitify.com	29.93	84	Not allowed	2.21	14/03/2012	8.40 am
Custom designed T-shirt, mug, calendar, etc.							
17	www.Zoomin.com						
18	www.Snapfish.in						
19	www.Picsquare.com	6.24	80	Not allowed	608.99	14/03/2012	8.50 am
Footwear							
20	www.Yebhi.com	44.07	83	Not allowed	1.8	14/03/2012	8.55 am
21	www.MetroShoes.com						
22	www.BataShoes.com						
23	www.BeStylish.com	4.23	85	Not allowed	1.44	14/03/2012	9.00 am

- **Internationalization**:
  - **Language**: The language defined in document and language used on the web pages must be same. And the same language should be used by the main target market [4].
- **Mobile compatibility**:
  - **Meta viewpoint**: It indicates how website should display on mobile devices. This tag allows web pages to fit on smaller screens. It is generally used in responsive designs [4].
  - **Heavy images**: Compressing or resizing the heaviest images in order to reduce their loading time is need of time. Try using a web-friendly format [4].
- **404**:
  - **Optimized 404**: The presence of an optimized 404 page allows avoiding the loss of visitors in the case of an unfound page. By customizing this page, one will be able to keep the same general style of web page and even include a search form to give visitors other possibilities [4].
- **Printing**:
  - **Printing style sheet**: Adding a dedicated style sheet for printing allows users to print the content of your website in a more readable way [4].

URL optimization parameter is not considered for our research work because the data did not support for the analysis. When this data was considered and processed by using Weka, it gave unexpected result. So this parameter is ignored in analysis.

3.3.2 Design

- **Standardization**:
  - **Doctype**: Indicating the doctype used in document is required to validate it. This doctype contains the Document Type Definition (DTD), which defines tagging standards used in web page [4].
  - **Meta charset**: It is important to define the set of characters used in order to reduce display problems [4].
  - **W3C**: An HTML document must be valid to be fully compatible and readable by browsers and robots [4].

- **Meta tags**:
  - **Meta Description**: It is recommended to have characters, which lie between 140 and 170. This meta tag should describe the page without being a straight forward list of keywords. On the other hand, it must contain the most important keyword expressions of the page [4].
  - **Meta Keywords**: This Meta tag allows inserting specific keywords in relation to the contents of the page. One can add several keywords by using a coma to separate them. Now-a-days, the importance of this Meta tag has been largely reduced if not totally excluded [4].
  - **Meta Robots**: This Meta tag indicates to search engines that they must index or not index, follow or not follow specific links. This tag can be responsible for the absence of website in the search engine's index [4].
  - **All Meta Tags**: Meta tags give indications on the page itself. This information does not appear in the visible/readable content of the web page [4].
- **Segmentation**:
  - **Header**: It contains all the elements of the header. Generally, these elements are the same on each page of the website. Usually, it contains the logo, the search bar, the page title, the menu, etc. [4].
  - **Nav**: It represents the main menu of the website and it contains the different links used for the navigation [4].
  - **Footer**: This tag contains all the elements of the footer like the TOS (Terms of Service), contact form, etc. This element cannot contain <nav> tags but can contain a secondary menu [4].
- **Design**:
  - **Cellular design**: It is recommended to use divs (The <div> tag defines a division or a section in an HTML document) for the structure of website [4].
- **CSS**:
  - **Inline CSS**: This technique is not recommended because it weighs more on the page. Instead of this, one should use CSS [4].
  - **Style Tags**: Try to avoid including styles directly in HTML file because this considerably increases the page weight. One can use this element for a couple of lines but no more than 2ko [4].
  - **Linked CSS**: It is recommended to combine CSS files in one single file to lower the number of HTTP requests when possible [4].

- **CSS file compression**: Compressing CSS files reduces their size and loading time [4].
- **Scripts**:
  - **Integrated scripts**: It is better to include Javascript in another separate file to optimize the loading time of the web page [4].
  - **Script compression**: Compressing scripts reduces their loading time [4].
  - **Scripts in footer**: It is recommended to put scripts at the end of document just before the </body> tag in order to improve the loading time [4]. The Table 3.3 shows sample of actual collected design data.

3.3.3 Texts

- **Text content**:
  - **Page title**: It is recommended that the page title includes characters that lie between 60 and 80 characters. The title of the page must contain the main keywords of the page and the name of the organization or the brand. This must not be used for keyword stuffing but to describe the content of the web page [4].
  - **Text/Code ratio**: The text/code ratio is used to determine the quality of a web page. When this ratio is higher, the contents are more important [4].
- **Semantic**:
  - **Titles**: The H (Heading) elements are used to organize the structure of the page in relation to their order of importance [4].
  - **Text styling**: It is very essential to highlight the key expressions of the page. Do not forget the difference between <strong> and <em> which are used to define the importance of certain expressions and the <b> and <i> tags which are used to create the graphic form of the text. These tags should not contain too many keywords in order to maintain their efficiency [4].
- **Keywords**:
  - **Title coherence**: Key expressions must be present in the title of the web page. If this is not the case, either page's title is not representative of its content or the keywords used in the content of the page are not correctly thought through [4].

Table 3.3 Actual collected design data

Sr. No.	Name of the URL	Doctype	Meta Charset	W3C (errors)	Meta Descrip-tion	Meta Key-words	Meta Robots	All Meta Tags	Cellular Design	In-line CSS	Style Tags	Linked CSS	CSS File Com-pres-sion	Integ-rated Scripts	Script Com-pres-sion	Scripts in the Footer	Date	Time
Fashion and Lifestyle																		
1	www.Sunglassesindia.com	XHTML 1.0	UTF-8	23	89	12	Missing	5	Not used	no	no	5	yes	no	yes	2	30/03/2012	1.40 pm
2	www.Brandsndeals.com	XHTML 1.0	Missing	103	Missing	absent	Missing	1	Not used	no	no	9	yes	no	yes	2	30/03/2012	1.45 pm
3	www.Majorbrands.in	HTML 5	UTF-8	232	Missing	absent	Missing	9	Not used	no	no	4	yes	no	yes	4	30/03/2012	1.50 pm
4	www.Pantaloons.com	HTML 5	UTF-8	77	159	Absent	Missing	4	Not used	no	no	14	yes	no	yes	1	30/03/2012	1.55 pm
5	www.Shoppersstop.com	XHTML 1.0	UTF-8	151	189	8	Missing	7	Not used	no	no	5	yes	no	yes	1	30/03/2012	2.00 pm
6	www.Inkfruit.com																	
7	www.UtsavSarees.com																	
8	www.Globus.com	Missing	Missing	12	Missing	Absent	Missing	0	Not used	no	no	no	no	no	no	no	30/03/2012	2.05 pm
9	www.zovi.com																	
10	www.BagsKart.com																	
11	www.Fetise.com	XHTML 1.0	UTF-8	19	212	8	Missing	10	Not used	no	no	16	yes	no	yes	8	30/03/2012	2.07 pm
12	www.Orosilber.com	XHTML 1.0	UTF-8	59	199	Absent	Missing	9	Not used	no	no	4	yes	no	yes	4	30/03/2012	2.09 pm
13	www.Violetbag.com	XHTML 1.1	UTF-8	205	117	9	Missing	10	Not used	no	no	2	yes	no	yes	3	30/03/2012	2.12 pm
14	www.Bags109.com																	
15	www.Myntra.com	HTML 5	Missing	65	175	17	Missing	22	Not used	no	no	4	yes	no	yes	no	30/03/2012	2.14 pm
Custom designed T-shirt, mug, calendar, etc.																		
17	www.Zoomin.com																	
18	www.Snapfish.in																	
19	www.Picsquare.com	XHTML 1.0	Missing	97	323	18	Missing	4	Not used	no	no	3	yes	no	yes	2	30/03/2012	2.18 pm
Footwear																		
20	www.Yebhi.com	HTML 5	Missing	67	148	60	Missing	14	Not used	no	no	4	yes	no	yes	3	30/03/2012	2.20 pm
21	www.MetroShoes.com																	
22	www.BataShoes.com																	
23	www.BeStylish.com	HTML 5	UTF-8	13	Missing	Absent	Missing	4	Not used	no	no	5	yes	no	yes	4	30/03/2012	2.22 pm

- **Keyword density**: Too many keywords decrease the importance given to the expressions. The number of repetitions of a keyword defines its density. The structure density takes into account the keyword's importance in the structure of the content [4].
- **Main keywords**: Need to concentrate on main keywords to target more precisely the main subject of content's page [4].

- **Security**:
  - **Scannable email addresses**: In security point of view, it is recommended to replace the email addresses by images or javascripts to avoid them being crawled by robots and used by spammers [4]. Table 3.4 shows sample of actual collected texts data.

3.3.4 Multimedia

- **Images**:
  - **Number of images**: It is recommended to use CSS sprites or javascript in order to load images quicker [4].
  - **Unreachable images**: Unreachable images return a 404 error code which means that they cannot be found at the specified address [4].
  - **Image caching**: Images do not often change. One should cache these images for at least a month in order to improve website's loading time [4].
  - **Alternative text**: The alt attribute in an image element informs the search engine of its description. Place keywords carefully. The length of the alternative text should be inferior to 80 characters. This attribute is also used by assistance software in order to describe the images for blind people [4].

- **Unrecommended content**:
  - **Flash content**: Flash animations have become obsolete and have been replaced by CSS3 and HTML5 animations. Flash animations are no longer taken into consideration by mobile devices [4].
  - **Frames and iFrames**: Frames and iFrames are not recommended because their content is not readable by search engines [4].

- **Icons**:
  - **App icons**: These icons are used as app icons on mobile devices [4].
  - **Favicon**: These favicons are used as bookmark icons in regular web browsers [4].

Table 3.4 Actual collected texts data

Sr. No.	Name of the URL	Page Title	Text/Code Ratio (%)	Titles						Text Styling			Title Coherence	Keyword Density (%)	Scannable Email Addresses	Time	Date
				H1	H2	H3	H4	H5	H6	Strong	U	EM					
Fashion and Lifestyle																	
1	www.Sunglassesindia.com	86	7.08	0	9	0	0	0	0	70	0	64	2 out of 11	18.48	Not found	8.25 am	31/03/2012
2	www.Brandsndeals.com	98	2.68	13	0	0	0	0	0	6	0	0	4 out of 12	60.81	Not found	8.28 am	31/03/2012
3	www.Majorbrands.in	89	3.37	2	7	0	0	0	0	0	0	0	2 out of 7	36.11	Not found	8.30 am	31/03/2012
4	www.Pantaloons.com	15	3.82	0	0	0	0	0	0	0	0	0	No keywords	26.77	Not found	8.32 am	31/03/2012
5	www.Shoppersstop.com	77	6.97	1	2	27	##	0	0	13	0	0	1 out of 11	14.52	Not found	8.33 am	31/03/2012
6	www.Inkfruit.com																
7	www.UtsavSarees.com																
8	www.Globus.com	24	26.6	0	0	0	0	0	0	0	0	0	4 out of 4	63.1	Not found	8.34 am	31/03/2012
9	www.zovi.com																
10	www.BagsKart.com																
11	www.Fetise.com	97	2.86	0	0	0	0	0	0	0	0	0	6 out of 12	68.69	Not found	8.36 am	31/03/2012
12	www.Orosilber.com	88	6.47	0	0	0	0	0	0	0	0	0	1 out of 9	63.17	Not found	8.37 am	31/03/2012
13	www.Violetbag.com	62	6.45	0	26	0	1	0	0	5	0	0	1 out of 8	36.59	Not found	8.38 am	31/03/2012
14	www.Bags109.com																
15	www.Myntra.com	85	1.53	0	0	0	0	0	0	0	0	0	1 out of 10	46.7	Not found	8.39 am	31/03/2012
Custom designed T-shirt, mug, calendar, etc.																	
17	www.Zoomin.com																
18	www.Snapfish.in																
19	www.Picsquare.com	128	12.07	1	6	6	0	12	0	0	0	12	4 out of 16	35.33	Not found	8.41 am	31/03/2012
Footwear																	
20	www.Yebhi.com	104	4.48	0	0	0	1	0	0	0	0	0	2 out of 13	37.33	Not found	8.42 am	31/03/2012
21	www.MetroShoes.com																
22	www.BataShoes.com																
23	www.BeStylish.com	101	4.56	0	0	0	20	0	0	0	0	0	1 out of 12	47.51	Not found	8.44 am	31/03/2012

Table 3.5 shows sample of actual collected multimedia data.

3.3.5 Networking

- **Links**:
  - **Number of links**: Too many links in a page reduces the importance of each one of them [4].
  - **Link juice**: It is the ratio between internal links and exiting links. It is used to recognize if the site is a gateway with the aim to redirect the user toward another site or not. It is recommended that this ratio must be more than 50% [4].
  - **Single links**: Having too many duplicate links on website is harmful for SEO. This score is obtained by measuring the ratio between single links and duplicate links [4].
  - **Links without underscores**: The use of underscores "_" in a link is not recommended because it is not recognized by search engines as word separators. Instead of underscore, one should use hyphens "-" [4].
  - **Title attributes**: The title attribute allows to associate keywords to the content of the specific page [4].
  - **Linking**: This attribute indicates to search engines whether or not this link should be followed. Too many "no-follow" links increase the distribution of the link juice and reduces the importance of the other links. Wherever possible avoid defining more than five "no-follow" links on the same page [4].
  - **Reliable links**: This ratio corresponds to the number of reliable links. That is to say, the number of links which search engines can crawl [4].
- **Rankings**:
  - **Alexa rank**: The statistics given by the private firm Alexa come from the Alexa toolbar installed on the browsers of millions of users across the world [4].
- **Duplicate content**:
  - **Canonical linking**: The attribute link rel="canonical" allows to avoid duplicate content. Inform the search engine of the preferred version of a page or content if there is more than one page with a similar content [4].

Table 3.5 Actual collected multimedia data

Sr. No.	Name of the URL	Number of Images	Unreach-able Images	Image Caching	Alternative Text (%)		Flash Content	Frames and iFrames	App Icons	Favicon	Time	Date
					POIWAT	PWCAT						
Fashion and Lifestyle												
1	www.Sunglassesindia.com	69	1	68	51.1	45.3	No	No	No	Found	10.38 am	31/03/2012
2	www.Brandsndeals.com	19	1	18	40	10	No	No	No	Found	10.40 am	31/03/2012
3	www.Majorbrands.in	24	No	1	8	8	No	No	No	Found	10.41 am	31/03/2012
4	www.Pantaloons.com	51	3	All	0	0	No	Yes	No	Found	11.11 am	31/03/2012
5	www.Shoppersstop.com	31	1	1	40.6	37.5	No	Yes	No	Found	11.10 am	31/03/2012
6	www.Inkfruit.com											
7	www.UtsavSarees.com											
8	www.Globus.com	No	No	No	0	0	No	No	No	Missing	11.14 am	31/03/2012
9	www.zovi.com											
10	www.BagsKart.com											
11	www.Fetise.com	18	1	18	52.2	17.4	No	Found	No	Missing	11.16 am	31/03/2012
12	www.Orosilber.com	28	No	3	20.7	17.2	No	Found	No	Found	11.17 am	31/03/2012
13	www.Violetbag.com	33	No	All	100	87.3	No	Found	No	Found	11.18 am	31/03/2012
14	www.Bags109.com											
15	www.Myntra.com	16	No	All	87.5	75	No	Found	No	Found	11.19 am	31/03/2012
16	www.Elitify.com	66	1	3	1.4	0	No	Found	No	Found	11.20 am	31/03/2012
Custom designed T-shirt, mug, calendar, etc.												
17	www.Zoomin.com											
18	www.Snapfish.in											
19	www.Picsquare.com	28	12	28	28.6	23.8	No	No	No	Missing	11.20 am	3/31/2012
Footwear												
20	www.Yebhi.com	16	No	14	93.8	31.3	No	No	No	Found	11.21 am	3/31/2012
21	www.MetroShoes.com											
22	www.BataShoes.com											
23	www.BeStylish.com	28	No	8	0	0	No	No	No	Found	11.22 am	3/31/2012

- **Indexation**:
  - **Robots.txt**: The robots.txt file allows you to authorize or forbid the indexation of a page or a directory by a search engine. It also permits to indicate the location of the sitemap(s) [4].
  - **Sitemap(s)**: A sitemap is a file containing an ordered organization of the linking structure of site. This file is used not only to define the importance attached to each page but also to help the search engines crawl the entire website. The location of the sitemap must be defined in the robots.txt file [4].
  - **DMoz**: The DMOZ open directory project is a trustworthy source used by search engines like Google to collect more information [4].
- **Social networks**:
  - **Facebook**: The meta property:"fb:page-id" can be used to link Facebook page to website. If linked correctly, information on the size of social network can be found there, otherwise it is blank [4].
  - **Twitter**: The information displayed here indicates the number of time website's URL has been shared on Twitter, allowing one to have an idea of the popularity of the website [4].
- **Search engines**:
  - **Search result preview**: This is a preview of website in the results of the search engines indexing [4].
  - **Last GoogleBot visit**: This date indicates the last time Google's robot crawled the page [4].
  - **Google indexed**: This number relates to the number of pages indexed in Google's search results [4].
  - **Bing indexed**: This number relates to the number of pages indexed in Bing's search results [4].
  - **Yahoo indexed**: This number relates to the number of pages indexed in Yahoo!'s search results [4].

Table 3.6 list out the values of the networking parameters. Actual values of all parameters have been recorded and stored in tables. The web parameter values are dynamic in nature. When all the actual values are recorded and stored in excel sheet for further processing, even downloading date and time are also recorded. All parameters values are recorded and stored for approximately 150 business websites.

Table 3.6 shows sample of actual collected networking data.

Table 3.6 Actual collected networking data

Sr. No.	Name of the URL	Number of Links	Link Juice (%)	Single Links %	Links without Underscores %	Title Attribute (%)	Linking (%)	Reliable Link	Date	Time
	Fashion and Lifestyle									
1	www.Sunglassesindia.com	270	3	68.5	75.9	0.7	100	100	13/03/2012	3.02 pm
2	www.Brandsndeals.com	56	14.3	78.6	98.2	37.5	100	91.1	13/03/2012	4.10 pm
3	www.Majorbrands.in	301	2.3	74.4	28.2	0.3	97	97	13/03/2012	4.12 pm
4	www.Pantaloons.com	241	69.7	41.5	97.9	0	99.6	72.6	13/03/2012	5.05 pm
5	www.Shoppersstop.com	1139	0.4	48.3	98.2	6	96.5	93.4	13/03/2012	5.11 pm
6	www.Inkfruit.com									
7	www.UtsavSarees.com									
8	www.Globus.com	1	100	100	100	0	100	100	13/03/2012	5.17 pm
9	www.Zovi.com									
10	www.BagsKart.com									
11	www.Fetise.com	59	25.4	71.2	67.8	11.9	100	91.5	13/03/2012	5.24 pm
12	www.Orosilber.com	80	1.3	86.2	91.2	6.3	100	88.8	13/03/2012	5.28 pm
13	www.Violetbag.com	278	0.7	94.6	67.3	1.1	97.1	97.5	13/03/2012	5.34 pm
14	www.Bags109.com									
15	www.Myntra.com	345	1.7	93.3	41.7	1.8	98.3	98.6	13/03/2012	5.38 pm
16	www.Elitify.com	679	1.2	63.8	95.9	40.2	100	94.8	14/03/2012	8.40 am
	Custom designed T-shirt, mug, calendar, etc.									
17	www.Zoomin.com									
18	www.Snapfish.in									
19	www.Picsquare.com	217	4.6	50.7	99.1	0	99.5	100	14/03/2012	8.50 am
	Footwear									
20	www.Yebhi.com	146	1.4	78.8	99.3	0	100	97.3	14/03/2012	8.55 am
21	www.MetroShoes.com									
22	www.BataShoes.com									
23	www.BeStylish.com	45	0	100	100	0	100	100	14/03/2012	9.00 am

3.4 Difficulties Encountered

3.4.1 Internet Problem

While collecting the data on the Internet, sometimes setting of proxy server poses some issues. Sometimes, the Internet connection with proxy gives results but the tool used for the collection of data is unable to display the result. At the same time, the tool takes the input and it takes too much time to display the result or no result at all. Sometimes, the speed is too slow to open the tool as well as processing is too slow with overburdening of data.

3.4.2 Preparation and Selection of Websites

For research work, approximately 150 commercial websites are selected. These websites are taken by searching on Google. There are so many commercial websites available but out of those randomly only approximately 150 websites are selected. The collected commercial websites are from different sectors such as footwear, cloths, hotels, tours, etc. There are so many websites for each category on Internet, but it was very difficult to choose few websites for the research work. For research work, at least three or four from each category are selected randomly. By using site-analyzer tool, the performance of these commercial websites was tested. The purpose to analyze the overall performance of the commercial websites is to attract more and more customers through which the organizations do more business and increase their profits. They can also prepare their own business strategies.

3.4.3 Difficulty in Selecting Analysis Tool

We were searching for the tool which can properly analyze the overall performance of the commercial websites. For that, the search was carried on Google search engine. There are so many tools that are available on web, but we selected the tool which is freely available. The site-analyzer tool which is freely downloadable and freely accessible. This tool was used for research work. We collected the data of selected commercial websites by using this tool.

3.4.4 Unavailability of Data

While collecting the data by using the site-analyzer tool, sometimes the result was displayed like 'Oops!' that means there is no data for the corresponding commercial website.

3.5 Flowchart

Figure 3.4 shows the flowchart of data collection.

Step 1: Starting the process

Step 2: Open Google to search for available tools on the Internet for website performance analysis. One tool available on the website http://www.site-analyzer.com for analysis of website [4] is site-analyzer.

Step 3: Input the name of the social network website to the tool.

Step 4: After providing the name of the social network website to the site-analyzer tool, it displays the values of different parameters such as global score, accessibility, design, texts, multimedia, and networking. The values of these

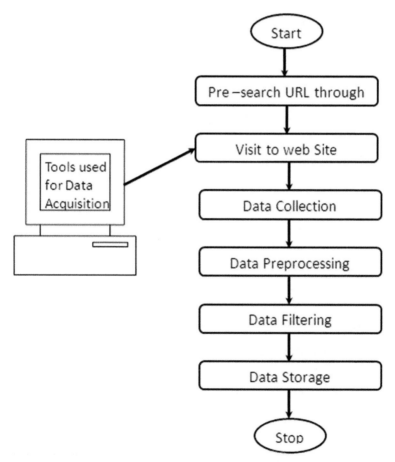

Figure 3.4   Flowchart for data collection.

parameters are displayed in percentage. While collecting these values of the parameters, we mention the time and date because the web content is dynamic. All these parameters are stored in excel sheet named Overall_Data.xlsx. At the same time, we collected the details of accessibility parameters such as page weight, compression, page caching, and download time. The values of these sub parameters are also recorded in Web_accessibility.xlsx file. Then, URL optimization detail values given by the site-analyzer tool are www redirection, URL rewriting, Domain length, domain age in years month days, and the last is Whois (gives Creation date, Expiration date, and Last update). These sub-parameter values of URL optimization are stored in excel sheet named URL_Optimization.xlsx. At the last, the Networking parameter's details such as Number of Links, Link Juice, Single links, Links without underscores, Title attribute, Linking, and Reliable links are recorded. These parameter values are stored in excel sheet named Networking.xlsx file.

Step 5: Once all the data of these parameters are collected of 123 social network websites, data preprocessing techniques are applied on the same for converting it to quality data. For few websites, the tool is unable to provide the values of the parameters, so such websites are not considered forever. In data preprocessing, deletion of unnecessary values of the parameters and retaining the valid values for further analysis are done.

Step 6: In data filtering, the fields are removed from the table that were not playing any active role in the processing and the results are not related/depend on such fields. For example, from each excel file the field name sr no is deleted because it will not affect the results.

Step 7: In the last step, data is stored in the excel file which is suitable for processing in Weka by giving extension either .arff (attribute relation file format) file or .csv (comma-separated values) file.

Step 8: Stop.

3.6 Freezing Parameters

3.6.1 Data Preprocessing

Most of the time, the real-world data is dirty. The data may contain so many errors, unexpected values, missing data, and duplication of data. After processing such types of dirty data, the result leads to wrong conclusion. It means that no quality data, so no quality mining results. For quality result, it is necessary to clean this dirty data. To convert dirty data into quality data, there is need for data processing techniques [1].

3.6.1.1 Data Preprocessing Techniques

There are number of data preprocessing techniques. Data cleaning can be applied on noisy and inconsistent data to make it cleaned one. Data integration combines data from multiple homogeneous or heterogeneous sources into a coherent data store, such as data warehouse. Data transformations may be applied on the data such as normalization, aggregation, generalization, and summarization. Data reduction reduces the size of the data by using aggregation and by eliminating redundant features [1]. In data cleaning, there are a number of techniques to handle missing values such as ignore the tuple, manually fill missing values, and fill it by automatically, etc. For present research work, first technique, i.e. ignore the tuple, is used.

In web accessibility dataset, there is one attribute named Page Caching having values 'allowed' and 'not allowed'. For present research work, we normalize these values to 1 and 0. Similarly, in URL optimization dataset, we normalize www redirection attribute to 0 for 'redirection is not active' and 1 for 'redirection is active'. Also, URL rewriting attribute 0 is applied for 'not used' and 1 is applied for 'used' values. The above activities are carried out under transformation phase of the data preprocessing techniques (Figure 3.5).

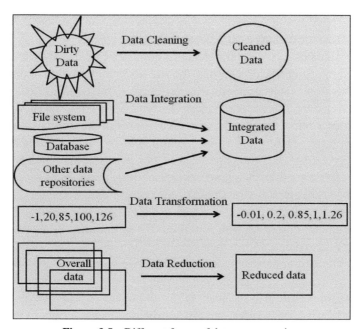

Figure 3.5 Different forms of data preprocessing.

3.6.2 Preprocessing and Filtering

3.6.2.1 Preprocessed and Filtered Overall Data

While collecting the data of the overall parameters, some websites data is given as null by the site-analyzer tool. All the values of parameters are null. From the actual collected overall data, these rows were deleted and the table is considered with the remaining commercial website parameter values given by the site-analyzer tool.

Table 3.7 shows sample of preprocessed and filtered overall data.

3.6.2.2 Preprocessed and Filtered Web Accessibility Data

The web accessibility data of commercial websites is collected by using the site-analyzer tool. For some commercial websites, the site-analyzer tool did not display the values. Therefore, such commercial websites are removed from the actual collected web accessibility data. For the parameters 'page caching', the site-analyzer gives the result 'not allowed' for all commercial websites. Hence, the values of page caching are replaced by zero and further used for processing.

Table 3.8 shows sample of preprocessed and filtered web accessibility data.

3.6.2.3 Preprocessed and Filtered Design Data

The Design data of commercial websites is collected by using the site-analyzer tool. For some commercial websites, the site-analyzer tool did not display the values. Therefore, such commercial websites are removed from the actual collected Design data. For the parameters Doctype, Meta Charset, W3C (errors), and Meta description, the site-analyzer gives the result 'missing' for some commercial websites. Hence, the values are replaced by blank space and further used for processing. For the parameter Meta keywords, the result is 'absent' for some commercial websites, so we replace it by zero. The next parameter Meta Robots have result 'missing' for all commercial websites. Therefore, this column is deleted from the actual Design data. The result of Cellular design parameter comes up with 'not used' for all commercial websites. This parameter is also deleted from data. Similarly, Inline CSS and Style tags have result 'no' for all commercial websites. These two parameters are also deleted from the collected design data. For Integrated scripts and Script compression parameters, the result is in the form of yes or no. We replace yes by 1 and no by 0 for analysis work. And at the last for the parameter scripts in the footer, some commercial websites come up with the numerical values or no values. Therefore, wherever no value is appearing in the data, we replace it by 0.

Table 3.7 Preprocessed and filtered overall data

Sr. No.	Name of the URL	Global Score	Accessibility	Design	Texts	Multimedia	Networking
1	www.Sunglassesindia.com	59.23	73.72	62.02	46.67	55.88	50.82
2	www.Brandsndeals.com	56.01	71.29	52.73	52.15	38.23	58.19
3	www.Majorbrands.in	55.91	56.39	66.67	52.03	55.88	42.62
4	www.Pantaloons.com	47.9	46.42	87.43	29.69	35.88	14.75
5	www.Shoppersstop.com	51.31	64.56	65.46	57.13	24.12	27.87
6	www.Globus.com	59.61	60.81	54.64	66.36	74.71	49.18
7	www.Fetise.com	44.62	49.46	59.29	45.4	17.06	35.25
8	www.Orosilber.com	53.47	57.91	62.39	38.77	24.12	68.03
9	www.Violetbag.com	61.23	74.63	64.25	48.51	61.18	51.64
10	www.Myntra.com	55.11	53.11	70.81	29.24	61.18	52.46
11	www.Elitify.com	57.22	64.87	75.96	46.28	24.12	52.46
12	www.Picsquare.com	51.66	60.71	52.73	61.08	31.18	45.08
13	www.Yebhi.com	56.68	48.95	76.5	41.26	63.53	45.08
14	www.BeStylish.com	61.81	64.87	72.13	38.46	55.88	67.21
15	www.Hushbabies.com	46.99	60.81	54.64	28.97	88.19	10.24
16	www.Firstcry.com	57.14	60.41	66.67	56.68	50	44.26
17	www.Babyoye.com	49.71	60.71	62.02	45.6	17.06	44.26
18	www.Hoopos.com	52.19	76.25	62.02	45.6	17.06	38.52
19	www.Wowkart.com	58.17	46.01	71.31	65.52	50	52.46
20	www.Babyproducts.co.in	62.56	50.68	75.96	53.73	89.41	45.9
21	www.Infibeam.com	67.58	79.9	86.89	51.98	43.53	54.1
22	www.IndianGiftsPortal.com	51.01	70.95	52.73	45.79	31.18	42.62
23	www.Talash.com	61.81	80.1	68.58	54.34	38.23	52.46
24	www.GlimGifts.com	57.58	55.75	62.02	48.5	88.19	43.03
25	www.Giftsandlifestyle.com	71.32	75	75.41	60.05	88.19	60.66

Table 3.8   Preprocessed and filtered web_accessibility data

Sr. No.	Name of the URL	Page Weight	Compression	Page Caching	Download Time
1	www.Sunglassesindia.com	12.32	87	0	684.17
2	www.Brandsndeals.com	11.59	65	0	1.26
3	www.Majorbrands.in	19.11	89	0	2.83
4	www.Pantaloons.com	11.69	80	0	6.3
5	www.Shoppersstop.com	30.97	87	0	2.4
6	www.Globus.com	2.34	64	0	2.34
7	www.Fetise.com	11.43	74	0	1.99
8	www.Orosilber.com	17.87	72	0	3.37
9	www.Violetbag.com	12.33	85	0	810.93
10	www.Myntra.com	136.64	51	0	1.77
11	www.Elitify.com	29.93	84	0	2.21
12	www.Picsquare.com	6.24	80	0	608.99
13	www.Yebhi.com	44.07	83	0	1.8
14	www.BeStylish.com	4.23	85	0	1.44
15	www.Hushbabies.com	4.22	55	0	4.22
16	www.Firstcry.com	60.99	73	0	662.13
17	www.Babyoye.com	25.15	84	0	1.7
18	www.Hoopos.com	25.17	84	0	1.7
19	www.Wowkart.com	111.62	87	0	4.87
20	www.Babyproducts.co.in	7.99	62	0	265.62
21	www.Infibeam.com	20.41	82	0	1.82
22	www.IndianGiftsPortal.com	38.88	88	0	850.47
23	www.Talash.com	15.34	78	0	1.89
24	www.GlimGifts.com	3.62	58	0	216.78
25	www.Giftsandlifestyle.com	3.43	62	0	390.49

All the values of the parameters are recorded on specific time and data, because the websites are dynamic in nature.

Table 3.9 shows sample of preprocessed and filtered design data.

3.6.2.4 Preprocessed and Filtered Texts Data

The Texts data of commercial websites is collected by using the site-analyzer tool. For some commercial websites, the site-analyzer tool did not display the values. Therefore, such commercial websites are removed from the actual collected Texts data. The parameter Text styling has three sub-parameters as Strong (strong tag defines strong emphasized text), U (underline), and EM (emphasis). For the Title coherence parameter site-analyzer gives the result like '2 out of 11', so for these parameters we take the ratio of these two values like 2/11. Sometimes for some commercial websites, site-analyzer gives the output like 'no keywords' so these can be replaced by 0. The ratio values are

Table 3.9 Preprocessed and filtered design data

Sr. No.	Name of the URL	Doctype	Meta Charset	W3C (Errors)	Meta Description	Meta Keywords	All Meta Tags	Linked CSS	CSS File Compression	Script Compression	Scripts in the Footer
1	www.Sunglassesindia.com	XHTML 1.0	UTF-8	23	89	12	5	5	1	1	2
2	www.Brandsndeals.com	XHTML 1.0		103		0	1	9	1	1	2
3	www.Majorbrands.in	HTML 5	UTF-8	232		0	9	4	1	1	4
4	www.Pantaloons.com	HTML 5	UTF-8	77	159	0	4	14	1	1	1
5	www.Shoppersstop.com	XHTML 1.0	UTF-8	151	189	8	7	5	1	1	1
6	www.Globus.com			12		0	0	0	0	0	0
7	www.Fetise.com	XHTML 1.0	UTF-8	19	212	8	10	16	1	1	8
8	www.Orosilber.com	XHTML 1.0	UTF-8	59	199	0	9	4	1	1	4
9	www.Violetbag.com	XHTML 1.1	UTF-8	205	117	9	10	2	1	1	3
10	www.Myntra.com	HTML 5		65	175	17	22	4	1	1	0
11	www.Elitify.com	HTML 5	UTF-8	504	167	18	25	4	1	1	2
12	www.Picsquare.com	XHTML 1.0		97	323	18	4	3	1	1	2
13	www.Yebhi.com	HTML 5		67	148	60	14	4	1	1	3
14	www.BeStylish.com	HTML 5	UTF-8	13		0	4	5	1	1	4
15	www.Hushbabies.com			11		0	2	0	0	0	0
16	www.Firstcry.com	HTML 5	UTF-8	137	241	7	6	2	1	1	4
17	www.Babyoye.com	XHTML 1.0	UTF-8	691	258	15	24	8	1	1	3
18	www.Hoopos.com	XHTML 1.0	UTF-8	691	258	15	24	8	1	1	3
19	www.Wowkart.com	XHTML 1.0	UTF-8	77	154	6	7	10	1	1	1
20	www.Babyproducts.co.in	HTML 5	UTF-8	1	150	0	6	0	0	1	3
21	www.Infibeam.com	HTML 5	UTF-8	63	165	15	9	3	1	1	0
22	www.IndianGiftsPortal.com	HTML 4.01		139		0	4	1	1	1	12
23	www.Talash.com	XHTML 1.0	UTF-8	81	151	18	8	12	1	1	7
24	www.GlimGifts.com	XHTML 1.0	UTF-8	25		0	1	4	1	1	2
25	www.Giftsandlifestyle.com	HTML5	UTF-8	0		0	3	8	1	1	4

used for the analysis work. For the next parameter, Scannable email addresses, the site-analyzer displays the output 'not found' for all commercial websites. So this column of Scannable email addresses is removed from the Texts data. The remaining rows and columns are used for further processing. All the values of the parameters are recorded on specific time and data because the websites are dynamic in nature.

Table 3.10 shows sample of preprocessed and filtered Texts data.

3.6.2.5 Preprocessed and Filtered Multimedia Data

The Multimedia data of commercial websites is collected by using the site-analyzer tool. For some commercial websites, the site-analyzer tool did not display the values. Therefore, those commercial websites are removed from the actual collected Multimedia data. For the Number of images parameter, site-analyzer gives the result as either any numeric value or 'no'. Therefore, we replaced 'no' as a –1. Similarly, for the Unreachable images parameter, site-analyzer gives the result as either any numeric value or 'no'. Found is treated as yes and then replaced 'no' as a –1. For the parameter Image Caching, the result is numeric value or 'all' or 'no', so we replaced 'all' to 100 and 'no' to –1. The Alternative text parameter has two sub-parameters as 'Percentage of images with alternative text' and 'Percentage with correct alternative text'. For these two sub-parameters, the output is either any numeric value or 'no', so 'no' can be replaced by –1. For the parameter Flash content, site-analyzer displays the result as 'no' for all commercial websites, so this parameter is deleted from the data. For the parameter Frames and iFrames, the site-analyzer gives the output as 'yes', 'no', and 'found'. Found is treated as yes and then replaced 'found' to 'yes' and then 'no' as 0 and 'yes' as 1. For almost all commercial websites, the site-analyzer displays the result 'no' for App icons' parameter. So, this parameter is deleted from the data. At the last, the parameter Favicon has either 'found' or 'missing' output given by the site-analyzer. So we replaced found to 0 and missing to 1. All the values of the parameters are recorded on specific time and date because of the dynamic nature of websites.

The Table 3.11 shows sample of preprocessed and filtered multimedia data.

3.6.2.6 Preprocessed and Filtered Networking Data

The Networking data of commercial websites is collected by using the site-analyzer tool. For some commercial websites, site-analyzer tool did not display the values. So such commercial websites are removed from the actual collected Networking data. All the values of the parameters are recorded on specific time and date, because the websites are dynamic in nature.

Table 3.10 Preprocessed and filtered texts data

Sr. No.	Name of the URL	Page Title	Text/Code Ratio	Titles						Text Styling			Title Coherence	Keyword Density
				H1	H2	H3	H4	H5	H6	Strong	U	EM		
1	www.Sunglassesindia.com	86	7.08	0	9	0	0	0	0	70	0	64	0.18	18.48
2	www.Brandsndeals.com	98	2.68	13	0	0	0	0	0	6	0	0	0.33	60.81
3	www.Majorbrands.in	89	3.37	2	7	0	0	0	0	0	0	0	0.29	36.11
4	www.Pantaloons.com	15	3.82	0	0	0	0	0	0	0	0	0	0	26.77
5	www.Shoppersstop.com	77	6.97	1	2	27	104	0	0	13	0	0	0.09	14.52
6	www.Globus.com	24	26.6	0	0	0	0	0	0	0	0	0	1.00	63.1
7	www.Fetise.com	97	2.86	0	0	0	0	0	0	0	0	0	0.50	68.69
8	www.Orosilber.com	88	6.47	0	0	0	0	0	0	0	0	0	0.11	63.17
9	www.Violetbag.com	62	6.45	0	26	0	1	0	0	5	0	0	0.13	36.59
10	www.Myntra.com	85	1.53	0	0	0	0	0	0	1	0	0	0.10	46.7
11	www.Elitify.com	76	7.4	0	1	0	0	1	4	1	0	0	0.10	36.06
12	www.Picsquare.com	128	12.07	1	6	6	0	12	0	0	0	12	0.25	35.33
13	www.Yebhi.com	104	4.48	0	0	0	1	0	0	0	0	0	0.15	37.33
14	www.BeStylish.com	101	4.56	0	0	0	20	0	0	0	0	0	0.08	47.51
15	www.Hushbabies.com	0	0	0	0	0	0	0	0	0	0	0	0	100
16	www.Firstcry.com	84	12.11	1	3	0	0	0	0	0	0	0	0.36	40.79
17	www.Babyoye.com	0	6.88	1	46	14	0	2	4	17	0	0	0	33.24
18	www.Hoopos.com	0	6.88	1	46	14	0	2	4	17	0	0	0	33.3
19	www.Wowkart.com	59	9.38	1	7	29	24	0	0	7	0	3	0.30	34.65
20	www.Babyproducts.co.in	65	0.98	1	2	0	0	0	0	0	0	0	0.60	73.08
21	www.Infibeam.com	88	7.21	1	0	13	0	0	0	15	0	0	0.09	29.46
22	www.IndianGiftsPortal.com	99	0.15	0	1	0	0	0	0	0	0	0	0.30	25.21
23	www.Talash.com	66	7.33	0	16	6	0	0	0	0	0	0	0.50	53.88
24	www.GlimGifts.com	13	4.52	0	0	0	0	0	0	0	0	0	1.00	88.1
25	www.Giftsandlifestyle.com	43	4.25	2	4	0	0	0	0	0	0	0	1.00	66.67

Table 3.11 Preprocessed and filtered multimedia data

Sr. No.	Name of the URL	Number of Images	Unreachable Images	Image Caching	Alternative Text		Frames and iFrames	Favicon
					POIWAT	PWCAT		
1	www.Sunglassesindia.com	69	1	68	51.1	45.3	0	0
2	www.Brandsndeals.com	19	1	18	40	10	0	0
3	www.Majorbrands.in	24	−1	1	8	8	0	0
4	www.Pantaloons.com	51	3	100	0	0	1	0
5	www.Shoppersstop.com	31	1	1	40.6	37.5	1	0
6	www.Globus.com	−1	−1	−1	0	0	0	1
7	www.Fetise.com	18	1	18	52.2	17.4	1	1
8	www.Orosilber.com	28	−1	3	20.7	17.2	1	0
9	www.Violetbag.com	33	−1	100	100	87.3	1	0
10	www.Myntra.com	16	−1	100	87.5	75	1	0
11	www.Elitify.com	66	−1	3	1.4	0	1	0
12	www.Picsquare.com	28	12	28	28.6	23.8	0	1
13	www.Yebhi.com	16	−1	14	93.8	31.3	0	0
14	www.BeStylish.com	28	−1	8	0	0	0	0
15	www.Hushbabies.com	−1	−1	−1	−1	−1	0	1
16	www.Firstcry.com	6	1	100	23.1	15.4	1	0
17	www.Babyoye.com	50	1	50	0.9	0	1	1
18	www.Hoopos.com	50	1	50	0.9	0	1	1
19	www.Wowkart.com	71	3	100	42.3	36.1	0	0
20	www.Babyproducts.co.in	−1	−1	−1	−1	−1	0	0
21	www.Infibeam.com	77	1	100	92.6	72.6	1	0
22	www.IndianGiftsPortal.com	3	2	3	1.6	1.6	1	1
23	www.Talash.com	37	2	7	53.7	43.9	1	0
24	www.GlimGifts.com	−1	−1	−1	−1	−1	0	1
25	www.Giftsandlifestyle.com	−1	−1	−1	−1	−1	0	1

Table 3.12 Preprocessed and filtered networking data

Sr. No.	Name of the URL	Number of Links	Link Juice	Single Links	Links without Underscores	Title Attribute	Linking	Reliable Link
1	www.Sunglassesindia.com	270	3	68.5	75.90	0.7	100	100
2	www.Brandsndeals.com	56	14.3	78.6	98.20	37.5	100	91.1
3	www.Majorbrands.in	301	2.3	74.4	28.20	0.3	97	97
4	www.Pantaloons.com	241	69.7	41.5	97.90	0	99.6	72.6
5	www.Shoppersstop.com	1139	0.4	48.3	98.20	6	96.5	93.4
6	www.Globus.com	1	100	100	100.00	0	100	100
7	www.Fetise.com	59	25.4	71.2	67.80	11.9	100	91.5
8	www.Orosilber.com	80	1.3	86.2	91.20	6.3	100	88.8
9	www.Violetbag.com	278	0.7	94.6	67.30	1.1	97.1	97.5
10	www.Myntra.com	345	1.7	93.3	41.70	1.8	98.3	98.6
11	www.Elitify.com	679	1.2	63.8	95.90	40.2	100	94.8
12	www.Picsquare.com	217	4.6	50.7	99.10	0	99.5	100
13	www.Yebhi.com	146	1.4	78.8	99.30	0	100	97.3
14	www.BeStylish.com	45	0	100	100.00	0	100	100
15	www.Firstcry.com	111	1.8	80.2	56.80	30.6	88.3	85.6
16	www.Babyoye.com	508	1	81.5	82.90	3.6	97.2	98.2
17	www.Hoopos.com	508	1.8	81.5	82.90	3.6	97.2	98.2
18	www.Wowkart.com	449	0.4	64.1	99.10	0	100	96.2
19	www.Babyproducts.co.in	6	83.3	33.3	0.00	0	100	100
20	www.Infibeam.com	236	4.2	57.2	83.10	37.3	96.6	97.9
21	www.IndianGiftsPortal.com	1682	0.1	73.1	99.90	0	99.9	82.2
22	www.Talash.com	284	1.4	87.30	99.30	6	100	94
23	www.GlimGifts.com	2	50	100	100.00	0	100	50
24	www.Giftsandlifestyle.com	12	100	83.3	100.00	8.3	100	100
25	www.Onamgifts.com	100	0	26	100.00	0	100	25

3.7 Way Forward

After successfully exemplifying the process of creation of dataset by taking a case study in web mining, the next chapter now takes you through the actual classification aspects of the dataset. Though the process of dataset creation pertains to e-commerce domain, the same methodology can be very well applied to any realm of big data.

Table 3.12 shows sample of preprocessed and filtered networking data.

References

[1] Tan, P.-N. Steinbach, M., and Kumar, V. (2005). *Introduction to Data Mining*. (Upper Saddle River, NJ: Pearson Education).

[2] Han, J., and Kamber, M. (2006). *Data Mining: Concepts and Techniques*, 2nd Edn. (Burlington, MA: Morgan Kaufmann Publishers, an imprint of Elsevier).

[3] Chen, H., and Chau, M. (2004). Web mining: machine learning for web applications. Ann. Rev. Inform. Sci. Technol. (ARIST) 38, 289–329.

[4] Website: http://www.site-analyzer.com accessed on 13, 14, 30 and 31 March, 2012.

4

Classification of Websites

Abstract

The initial part of the book focused on the evolution of web mining from the traditional data mining paradigm. Now it is time for the real application, and the case study covered in this chapter is classification of websites on the basis of standard metrics in order to help the business houses to have Search Engine Optimized (SEO) sticky websites in place. With the growing realm of electronic transactions, the need of such websites is undisputed. The dataset aspects for these websites have already been placed in Chapter 3. The attributes taken for classification are like accessibility, design, texts, multimedia, and networking. These attributes helps in predicting the SEO compatibility of the websites. The best part of this chapter is the actual screen shot and code written in Weka for the exercise of classification using different algorithms, namely J48 (Decision tree-based algorithm), RBFNetwork (neural network based), NaiveBayes (statistical method) and SMO (Support vector machine based). A thorough analysis of the classification vis-à-vis performance of the algorithm is dealt in depth. The know-how carved out of this chapter is not only significant from the web mining point of view, but it is vital in the web usability analysis.

4.1 Introduction

This is the era of e-marketing. Marketing is one of the ways to reach the likely customers. In this digital world, most of the business houses have their own websites. Creating website is not enough, as there are thousands of websites that already exist. For websites to help in business, they should reach to netizens. One of the most popular ways to reach the netizens is through search engines. Only creating and hosting the website is not sufficient, but it should be search engine optimized. Search Engine Optimization (SEO) is the technique which can help the websites by increasing the number of visitors to it, increasing the browsing time of each visitor and high-ranking placement in the search result of search engines. This ranking by search engine is based on what is highly relevant to user search keywords. Once the website appears on the first page of the search engine result page, then it will be automatically visited by most of the users (its human tendency).

Overall goal of most of the business websites is to appear on the first page of search engine result. There are many parameters associated with the websites to be analyzed for SEO compatibility. In fact, SEO is one of the marketing techniques for websites. As already mentioned in Chapter 3, a site-analyzer tool is used for SEO analysis of the websites. This analysis also helps in increasing the overall performance of the websites by identifying the loop holes and thus increasing the number of visitors to the website. More visitors do more business. The result of this analysis is stored as a dataset for classification. This tool analyzes the websites using the following benchmarks.

4.1.1 Accessibility

Here, accessibility is quick and easy approach to the website. From the user point of view, the website should have simple and good interface, should have very less downloading time, page weight, there should not be many redirecting links, etc. Other factors like making website content should be interesting and relevant and also play an important role. From search engine point of view, increasing its visibility to search engines will be increasing its rank.

4.1.2 Design

Web designing techniques are flourishing the market nowadays. They have been evolving since the last two decades and they started from simple HTML static pages to CSS and Scripts and many more. A design technique which is emerging is responsive web design. Mobile Internet access has increased tremendously since the last many years. There is need for web design that can change depending on the device accessing the website. Website structure should be adjustable and should give good resolution even when used on cell phones or tablets. This type of web designing is called responsive design. Responsive sites are more popular as easily available to the users through the cell phones. Users need the webpages which are easily navigable on their PCs, smart phones, or tablets if a webpage does not have good access or if not visible properly, then user will navigate away from such webpages. Hence, it takes time for business websites to be responsive.

Owing to the adaptability of responsive web design, there is no need to create separate websites for cell phones, which automatically increases the SEO of the website. Developing a separate website for different devices will have different URLs. Here, there is only one website with different URs, thus decreasing the SEO of the website as search engine rank will be different for different URLs. Thus the ranking of the website will be decreased. Other factors may be the use of CSS, embedded scripts, meta tags, etc. There is a need of business owners to have their websites with good usability, appearance, and ranking, so SEO and web designing have to go together.

4.1.3 Texts

A webpage should have good and informative text on it. From user point of view, the webpage should not be loaded with text, keywords can be highlighted, should provide easy navigation, hyper links, title tags, anchor text, etc. From SEO point of view, Anchor text, invisible text, and text to code ratio plays an important role.

Anchor text is the tools to name the links with some attractive and short labels. Usually, all the links have a URL address attached to it which one may not want to read and click, using anchor text these links can be given a short and informative name which describes the link. Thus, making it user friendly as user will be interested in clicking anchor text rather than URL text. Backlinks to the websites make the websites valuable from search engine point of view. Anchor text on the link give the search engine a fast and dependable way of knowing the keywords targeted by the webpage. This will increase the rank of the page and will also easy indexing by the search engines.

Invisible text as the name suggests is not visible to the users or visitors of the website. Here, the text is mixed in the background of the website. Invisible text is used for higher ranking by repeating the keywords or phrases for which the page should be ranked by the search engine. This repetition can be in the html code or at the footer blended with the background.

Text-to-code ratio is the ratio of front-end text appearing on the webpage to the html code behind the page. Text on the page represents its content. This content should be informative as there are content-based searches carried out by some search engines like Google panda. There ranking is content based. Another advantage of good text to code ratio is that it increases the loading speed of the page as the code is small. Users are happy with the speed and also it is considered by search engines in ranking the webpage. More plain text provides search engines the easy way to crawl and index the website.

4.1.4 Multimedia

It is all about pictures, animations, sound files, videos, etc., on a webpage. When multimedia is used on the webpage, concerns are: accessibility to wide range of website visitors, download time, appropriate player to play, slow connections, etc. Multimedia plays an important role in delivering the information in easy way and also provides easy access to physically challenged.

To make multimedia SEO friendly sounds difficult, as till now everything was concerned with text. But it is very easy as text associated with the multimedia in the form of filenames, tags, captions, etc. In short, text surrounding the multimedia can be made SEO friendly to increase the page rank. Using keyword-based names rather than generic names for images will make it SEO friendly. ALT tags for the images should be specified, which is also used by Google for its images. Content around the images plays an important role in categorizing the image for context and relevance, which is further used for indexing by search engines as they cannot analyze an image still. Filenames for the multimedia should also be content specific as search engines also index multimedia file names.

4.1.5 Networking

Networking is a network of Links. It deals with two types of links external and internal. Internal links are the connections of webpages of the same website, whereas external links are connection of webpages to another or different websites. External links tell what others say about you and internal links tell what you say about yourself. Links are like streets between webpages.

A website having links from other websites is called external links. These links point to external domain and are also links coming from external domain. External links are important as they pass link juice and are one of the parameters considered by search engines in ranking the webpage. In fact, external links decide the popularity of the webpage.

A website's links having links to its own pages are internal links. These links help in building the website structure. These structures help the web crawlers by providing them paths to crawl. Availability of main link navigation access to search engines increases page's search engine index.

Considering all the above parameters, an attempt to classify websites with base as SEO has been carried out. Each attribute is analyzed separately for classification. Overall analysis considering all the parameters at once is also carried out. Following sections deals with the same.

4.2 Classification of Websites on Accessibility

Accessibility deals with user as well as search engine interface. It is a tool to provide platform to the netizens to easily move around the website and interact with it. This is possible if a website has well labeled links and multimedia, user-friendly design which should be attractive and content rich so that users will spend more time consuming the contents on the website. This will automatically make the website popular and is an achievement from business point of view.

4.2.1 Dataset

A dataset with 123 instances is created as mentioned in Chapter 3 with all the attributes related to accessibility. This dataset is classified to find out the websites which provide Fast, moderate, or slow accessibility. Here, the classification is carried out using clustering approach. Before carrying out the clustering, data filtering was carried out. Here, attributes which do not play any specific role in clustering were filtered like sr. no., URL name, and page caching. Page caching was allowed by only one website in the dataset so this attribute was filtered.

4.2.2 Clustering

Clustering is carried out using attributes which are selected using select attribute menu of Weka. Each attribute was evaluated using attribute evaluator such as CfsSubsetEval, ClassifierSubsetEval, FilteredAttributeEval, etc. Attributes which were selected using attribute menu are as follows:

- **Page weight attribute**: Approximate page weight is taken which consists of text, animation, graphics, images, etc., on the webpage. If the weight is high, it will take time to download. Users will not wait for the webpage to download for minutes. Even if the website is good, it is not easily accessible by the users especially for slow connections.
- **Compression attribute**: It gives the percentage of compression supported by the website server. Most of the web browsers and web servers support http compression. HTTP Compression is a technique supported by most web browsers and web servers. If this is enabled on web browsers

as well as web servers, it will automatically reduce the size of text downloads (including HTML, CSS, and JavaScript) by 50–90%. Having compression-enabled websites will be beneficial to users and will also have potential saving in bandwidth costs at the server side.

- **Download Time**: Users are happy with the websites where webpages download quickly. There are lots of efforts spent by the web designers to speed up the download time of a webpage. This is an important attribute in website popularization as users are turned off by slow-loading pages.

The above three attributes are actually used in clustering the websites based on accessibility.

As data is numeric in nature, simple k-Means algorithm is used for clustering. Clustering of dataset is carried out using open source data mining tool Weka. Accessibility dataset is clustered using default settings of k-Means. It generated two Clusters. Clusters generated were not good as the sum of squared error was approximately 10.0234#. This indicated data cannot be clustered into two categories, so decision was taken to cluster data into three categories and surprisingly three good clusters were generated. Clusters were good as the sum of squared error was negligible. Following is the distribution of 123 instances into three clusters.

Here is the result of clustering with Weka.

```
****************************************************************
=== Run information ===
Scheme:weka.clusterers.SimpleKMeans  -N  3  -A  "weka.core.Euclidean
Distance -R first-last" -I 500 -S 10
Relation: accessibility-weka.filters.unsupervised.attribute.Remove-R1-2,5
Instances: 123

Attributes: 3
     Page Weight
     Compression
     Download Time
Test mode: evaluate on training data
=== Model and evaluation on training set ===
kMeans
======
Number of iterations: 3
Within cluster sum of squared errors: 0.007654339864911
Missing values globally replaced with mean/mode
Cluster centroids:
```

Attribute	Full Data	Cluster# 0	1	2
	(123)	(13)	(89)	(21)
=======	=======	=======	=======	=======
Page Weight	74.919	208.8292	67.3247	24.2076
Compression	72.5935	72.3846	70.4719	81.7143
Download Time	265.9386	908.6908	83.2578	642.2633
************	************	************	************	************

4.2.3 Clustered Instances

Output of Weka clustering is stored in result.arff file. In Weka, cluster number starts with zero but for analysis they are renumbered from one.

Cluster-1   13 (11%)
Cluster-2   89 (72%)
Cluster-3   21 (17%)

Features of websites in Cluster-1:
Intra-cluster analysis is carried out. Analysis showed that Websites in this cluster are slow to download as page weight is more and another reason for slow download is compression percentage, which is approximately 72%. Hence, the class Label for this cluster is decided as Slow. All the websites belonging to this cluster are slow in accessibility.

Features of websites in Cluster-2:
Here, download time of webpages is very less when compared with cluster-1. This is one of the factors in popularization of websites among netizens. Compression percentage is bit less than cluster-1 websites, but it does not affect here much as page weight is quite less. Owing to these factors, the download time is minimum compared with both the clusters. All these factors make websites in this cluster as websites with fast accessibility.

Features of websites in Cluster-3:
Intra-cluster analysis of this cluster puts it into the category of moderate accessibility websites. Here, page weight is minimum compared with other clusters, and compression ratio is also quite good, but the download time is very slow when compared with cluster-1. Looking at all these factors, this cluster is labeled as moderate accessibility cluster.

4.2.4 Classification Via Clustering

After clustering the dataset, we could get the class labels for each cluster. result.arff is updated now by replacing the cluster number attribute with class.

As per the features of each cluster class, labels are Fast, Moderate, and Slow. Here, class label is attached to all the instances of the dataset. Websites in this dataset will be classified as Fast, Moderate, and Slow accessible websites.

Here is the part of result.arff file with class labels.
```
***********************************************************
42,18.97,84,916.71,Slow
43,14.15,77,1.23,Fast
44,11.43,75,725.62,Moderate
45,58.13,87,1002.42,Slow
46,41.25,76,476.35,Moderate
***********************************************************
```

4.2.4.1 Classification via clustering using J48 algorithm

This updated result.arff file instances are classified using Weka J48 classifier. Tree classifier is used as decision trees can be converted into a set of rules easily. They can handle both numeric as well as nominal attributes which result.arff has. Another reason for using decision tree is they are nonparametric and make no assumptions about the space distribution and the classifier structure. Figure 4.1 shows the output of executing J48 Weka classification algorithm on result.arff.

Confusion matrix shows that except one instance, all the dataset is classified correctly.
```
***********************************************************
=== Confusion Matrix ===
 a    b    c      <- classified as
13    0    0  |   a = Slow
 0   88    1  |   b = Fast
 0    0   21  |   c = Moderate
***********************************************************
```

As per Graph 4.1, it can be observed that, out of 123 websites, 88 provide good accessibility, which is a plus point from business point of view. www.Myntra.com, www.Redbus.in, www.jetliteindia.co.in, www.flipkart.com, etc., belong to this category.

Out of 123 websites, 21 websites provide an average accessibility and need to decrease their download time. www.IndiaTimes.com, www.fashionara.com, www.Indiandailydeals.com, www.Dell.co.in, etc., belong to this category.

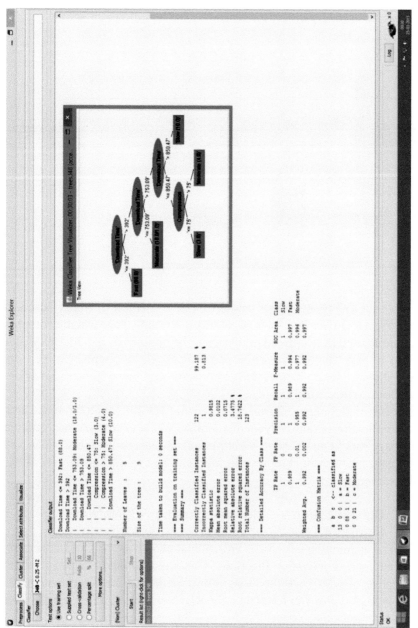

Figure 4.1 Classification based on accessibility using J48 algorithm.

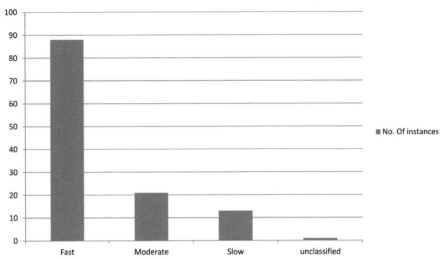

Graph 4.1   Classified websites.

There are 13 websites which belong to slow accessibility category and need to improve a lot to increase their popularity. Especially page weight, compression ratio as well as download time need to be optimized. It is like a red alert to these websites to survive in the competitive world. www.ClearTrip.com, www.indianairlines.com, www.Dealsonline4u.com, www.Homeshop18.com, etc., belong to this category.

Dataset used for analysis was collected on 13/3/2012 at 4:00 pm onwards. This classification is valid only for this data. As Websites are dynamic in nature, where page weight may change over time and thus affecting the download time of webpages, and even compression ratio can be changed at the web server by the administrator.

4.2.4.2 Classification via clustering using RBFNetwork algorithm

A radial basis function network is an artificial neural network that uses radial basis functions as activation functions. The output of the network is a linear combination of radial basis functions of the inputs and neuron parameters. The dataset is loaded into Weka and then the RBFNetwork classifier is fired on it. Figure 4.2 shows the output of executing RBFNetwork Weka classification algorithm on result.arff.

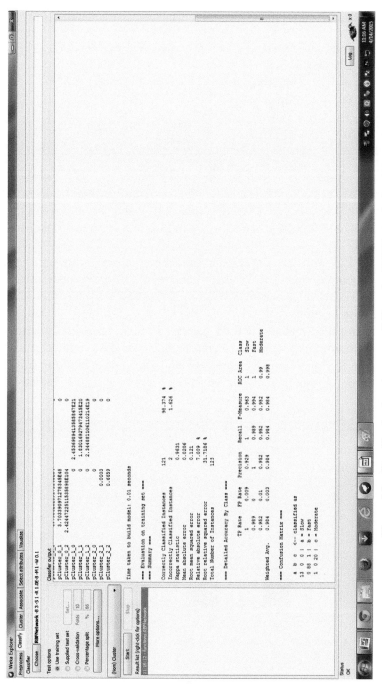

Figure 4.2 Classification based on accessibility using RBFNetwork algorithm.

Confusion matrix shows that except two instances all the dataset is classified correctly.

****

=== Confusion Matrix ===

```
a    b    c     <- classified as
13   0    0  |  a = Slow
0    88   1  |  b = Fast
1    0    20 |  c = Moderate
```

****

As per Graph 4.2, it can be observed that, out of 123 websites, 88 of them provide good accessibility, which is fruitful from business point of view. The websites that come under this category are www.Talash.com, www.GlimGifts.com, www.Giftsandlifestyle.com, www.MakeMyTrip.com, etc.

Out of 123 websites, 20 provide an average accessibility and need to decrease their download time. The websites that come under this category are www.Travelguru.com, www.SnapDeal.com, www.Taxiforsure.com, www.amazon.in, etc.

There are 13 websites, which belong to slow accessibility category and need to improve a lot to increase their popularity. It is necessary to optimize the page weight, compression ratio, and download time. As a competitor in the today's world, it is very essential to increase the

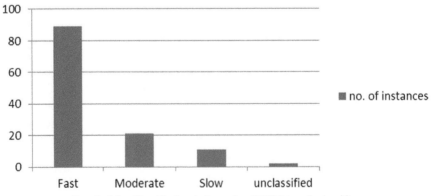

Graph 4.2   Classified websites using RBFNetwork classifier.

accessibility of their websites. The websites that come under this category are www.Spectglasses.com, www.Caratlane.com, www.Dealsonline4u.com, www.Goodlife.com, www.Jabong.com, www.Junglee.com, www.Homeshop 18.com, www.MakeMyTrip.com, etc.

Dataset used for analysis was collected on 13/03/2012 at 4.00 pm onwards. This classification is valid only for this data. Websites are dynamic in nature, where page weight may change over time, affecting download time of webpages, and even compression ratio can be changed at the web server by the administrator.

4.2.4.3 Classification via clustering using NaiveBayes algorithm

NaiveBayes classifiers are highly scalable, requiring a number of parameters linear in the number of variables (features/predictors) in a learning problem. The dataset is loaded into Weka and then by using NaiveBayes classifier, it is classified into three distinct classes. Figure 4.3 shows the output of executing NaiveBayes Weka classification algorithm on result.arff.

Confusion matrix shows that 117 instances are classified correctly and only six instances are classified incorrectly.

```
**************************************************************
=== Confusion Matrix ===
a    b    c      <– classified as
7    0    6  |   a = Slow
0   89    0  |   b = Fast
0    0   21  |   c = Moderate
**************************************************************
```

As per Graph 4.3, it can be observed that, out of 123 websites, 89 provide good accessibility. In business point of view, it is creditable. The websites that come under this category are www.Shoppersstop.com, www.Babyoye.com, www.BrandsKnot.com, www.99labels.com, www.Housefull.co.in, etc.

Out of 123 websites, 21 provide an average accessibility and need to decrease their download time. The websites that come under this category are www.Sunglassesindia.com, www.Violetbag.com, www.Picsquare.com, www.Firstcry.com, www.TravelMasti.com, www.IndiaTimes.com, www.Snap Deal.com, www.Dealbharat.in, www.Groffr.com, www.fashionara.com, etc.

There are 7 websites, which belong to slow accessibility category and need to improve a lot to increase their popularity. It is necessary to

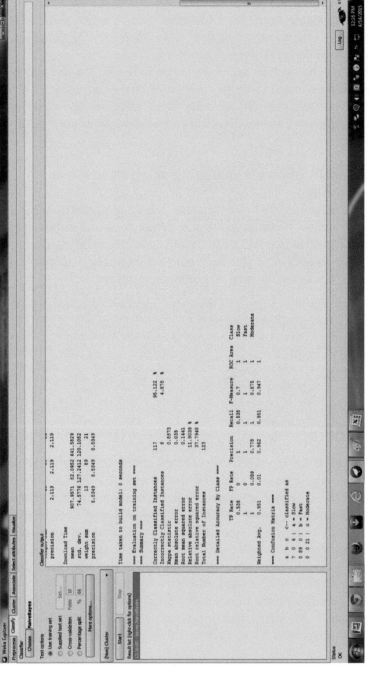

Figure 4.3   Classification based on accessibility using NaiveBayes algorithm.

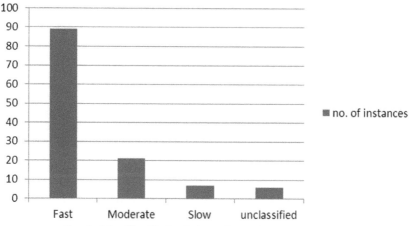

Graph 4.3 Classified websites using NaiveBayes classifier.

optimize the page weight, compression ratio, and download time. As a competitor in the today's world, it is very essential to increase the accessibility of their websites. The websites that come under this category are www.ClearTrip.com, www.Via.com, www.Goodlife.com, www.Junglee.com, www.Spectglasses.com, etc.

Dataset used for analysis was collected on 13/03/2012 at 4.00 pm onwards. This classification is valid only for this data. Websites are dynamic in nature, where page weight may change over time, affecting the download time of webpages, and even compression ratio can be changed at the web server by the administrator.

4.2.4.4 Classification via clustering using SMO algorithm

In machine learning, support vector machines are supervised learning models with associated learning algorithms that analyze data and recognize patterns, used for classification and regression analysis. It is used to classify linear or nonlinear data. Sequential Minimal Optimization (SMO), a fast method to solve huge quadratic programming problems, is widely used to speed up the training of the Support Vector Machines (SVMs). The dataset is loaded into Weka and, then by using SMO classifier, it is classified into three distinct classes. Figure 4.4 shows the output of executing SMO Weka classification algorithm on result.arff.

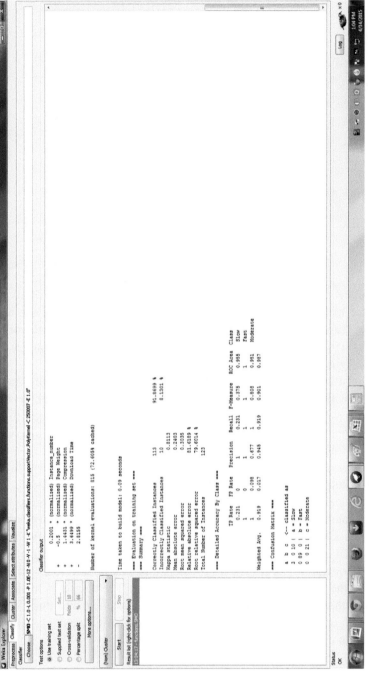

Figure 4.4   Classification based on accessibility using SMO algorithm.

Confusion matrix shows that 113 instances are classified correctly and only ten instances are classified incorrectly.

```
******************************************************************
=== Confusion Matrix ===
a    b    c      <- classified as
3    0    10  |   a = Slow
0    89   0   |   b = Fast
0    0    21  |   c = Moderate
******************************************************************
```

As per Graph 4.4, it can be observed that, out of 123 websites, 89 provide good accessibility. In business point of view, it is creditable. The websites that come under this category are www.Brandsndeals.com, www.Yebhi.com, www.Goair.in, www.Ezone.com, www.Machpowertools.com, www.Housefull.co.in, etc.

Out of 123 websites, 21 provide an average accessibility and need to decrease their download time. The websites that come under this category are www.fashionara.com, www.Taxiforsure.com, www. Snapittoday.com, www.LetsBuyProducts.com, www.SnapDeal.com, www.IndianGiftsPortal.com, www.Violetbag.com, etc.

There are three websites which belong to slow accessibility category and need to improve a lot to increase their popularity. It is necessary to optimize the page weight, compression ratio, and download time. As a competitor in the today's world, it is very essential to increase the accessibility of their websites.

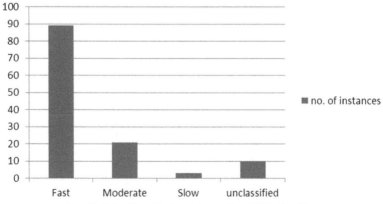

Graph 4.4 Classified websites using SMO classifier.

The websites that come under this category are www.indianairlines.com, www.Dealsonline4u.com, www.Spectglasses.com, etc.

Dataset used for analysis was collected on 13/03/2012 at 4.00 pm onwards. This classification is valid only for this data. As websites are dynamic in nature, where page weight may change over time, it affects the download time of webpages and even compression ratio can be changed at the web server by the administrator.

4.2.4.5 Comparison of above classification algorithms

The above classification algorithms display the results in Tables 4.1 and 4.2, Graphs 4.5 and 4.6. Table 4.1 and Graph 4.5 display how 123 instances are classified correctly and incorrectly by the four classification algorithms supported by Weka. The time required to process the dataset is also given in seconds. From Table 4.1 and Graph 4.5, it is observed that the accuracy of the J48 algorithm is more when compared with the remaining three algorithms, and the time required is also less. So, the J48 algorithm outperforms for the classification of dataset of the commercial websites.

From Table 4.2 and Graph 4.6, it is observed that kappa statistics value of J48 is more accurate as compared with other algorithms and, at the same time, other error values are less as compared with other algorithms. So, J48 is more suitable for classification of dataset of the business websites.

4.3 Classification Based on Website Design

Website design plays a major role in SEO. There are many factors associated like the structure of the websites, link map, site map, technology used in designing, user interface, browser and search engine friendly, site loading time, Meta tags used for easy navigation and search, the site as per the norms of W3C etc. Some of these features are used in creation of dataset through the site-analyzer tool for analysis.

Table 4.1 Correctly and incorrectly classified instances

Sr. No.	Algorithm Used	Correctly Classified Instances	% of Correctly Classified Instance	Incorrectly Classified Instances	% of incorrectly Classified Instance	Time in Sec.
1	J48	122	99.187	1	0.813	0
2	RBFNetwork	121	98.374	2	1.626	0.01
3	NaiveBayes	117	95.122	6	4.878	0
4	SMO	113	91.8699	10	8.1301	0.03

Table 4.2 Errors given by classification algorithms

Sr. No.	Algorithm Used	Kappa Statistic	Mean Absolute Error	Root Mean Squared Error	Relative Absolute Error	Root Relative Squared Error
1	J48	0.9815	0.0102	0.0715	3.4775	18.7622
2	RBFNetwork	0.9626	0.012	0.1015	4.0865	26.6044
3	NaiveBayes	0.8873	0.035	0.1441	11.9039	37.7948
4	SMO	0.8113	0.2403	0.3035	81.6189	79.6014

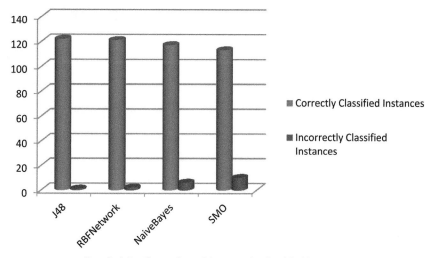

Graph 4.5 Correctly and incorrectly classified instances.

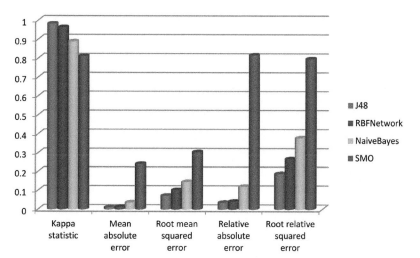

Graph 4.6 Errors given by classification algorithms.

The dataset prepared is design_analysis.csv, so that it can be directly used with Weka. Dataset consists of 122 instances with lots of missing values as data was not available for few attributes. The dataset has the following attributes:

Doctype (Document Type): It is a declaration for the type of the document webpage being used. There many different types of document that exist like HTML5.0, HTML4.1, XHTML1.0, etc. Owing to different types of coding styles, it is not easy for the browser to display the webpage properly, as it needs to parse the whole page or make a guess to check its compatibility and thus reduces the speed of loading and also accessibility. Here, this problem is solved by mentioning the doctype for the webpages, thus helping the browser in deciding the standard to be used for parsing the contents.

Meta charset: It is a type of characters used for coding the websites. The charset used by the website should be informed to the browser for correct display of data. There are two types here UTF-8 and ISO-8859-1. From SEO point of view, it does not make difference as to which character set is used but UTF-8 supports more characters as compared with ISO-8859-1.

W3C (errors): W3C is a World Wide Web consortium, which develops the web standards using which all the websites are tested. These errors should be minimum for SEO compatibility, as poor code quality can decrease the ranking of the website.

Meta description: As the name suggests, it is used to summarize the webpage content. It may not be useful for search engine ranking but can help in getting more user clicks if the webpage appears in search engine output. Looking at the description of the webpage, the user can decide whether to click or not. So, giving a good meta description indirectly increases the popularity of the webpages and hence are important. This description should be short not more than 155 characters approximately.

Meta Keywords: Unlike Meta description, meta tag do not appear as part of search engine result. Most of the search engines like Google and Bing do not use it for ranking but Yahoo and some other small search engines still use these Meta keywords for ranking. These keywords should be unique and should use synonyms.

All Meta tags: It is a count of different tags that exist on the webpage. There are large numbers of meta tags, which can be used but very few of them are SEO friendly.

Linked CSS (Cascading Style Sheets): CSS helps in attaching different styles to the webpages like color, spaces, fonts, etc. It also helps in controlling the page appearance in a flexible way. An important role of CSS is that they make the website SEO friendly. If a website is fully CSS functioning, it goes easy for the web crawlers to index and thus increase the rank of the webpage.

CSS file compression: This attribute has binary value if the CSS files have a compressed value of 1, otherwise 0. Compression is directly associated with the page speed as small size pages get downloaded faster and increase the accessibility. The page speed is one of the factors counted by search engines in ranking a webpage.

Script compression: This attribute is similar to CSS file compression with a binary value. Here, scripts in the code are compressed to speed the site loading time.

Scripts in the footer: As per the survey, these scripts carried out do not play any role in the SEO compatibility and are sometimes ignored by the search engines for the ranking.

4.3.1 Attribute Selection

Above dataset is submitted to Weka for attribute selection as all attributes may not play a vital role in classification. There are many selection techniques like CfsSubsetEval, ClassifierSubsetEval, Filtered AttributeEval, etc. Figure 4.5 shows the output of one of the evaluators.

After applying the above-mentioned feature, selection techniques and expert advice following four attributes were selected for classification:

- W3C(errors)
- Linked CSS
- CSS file compression
- Script compression.

4.3.2 Clustering

Once the attributes are freezed for classification, the same were used for clustering. Dataset is clustered using Weka with simple k-means as clustering technique. Properties of clustering technique were changed to get three clusters, as default is two. Figure 4.6 shows the output of clustering.

```
**************************************************************
=== Run information ===
Scheme:weka.clusterers.SimpleKMeans  -N  3  -A  "weka.core.Euclidean
Distance -R first-last" -I 500 -S 10
Relation:   Design   _analysis2-weka.filters.unsupervised.attribute.Remove-
R1-4,6-8-weka.filters.unsupervised.attribute.Remove-R5
Instances: 120
```

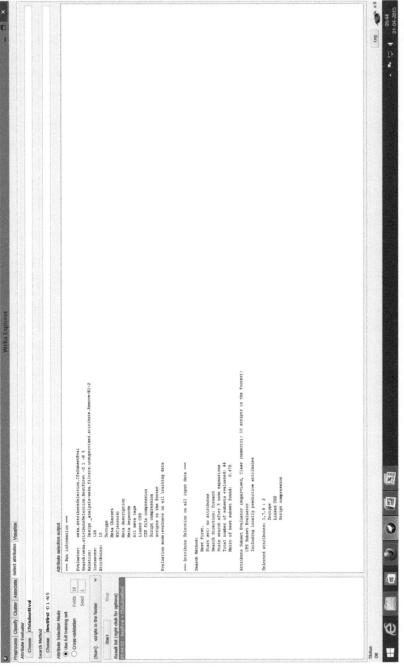

Figure 4.5   Attribute selection.

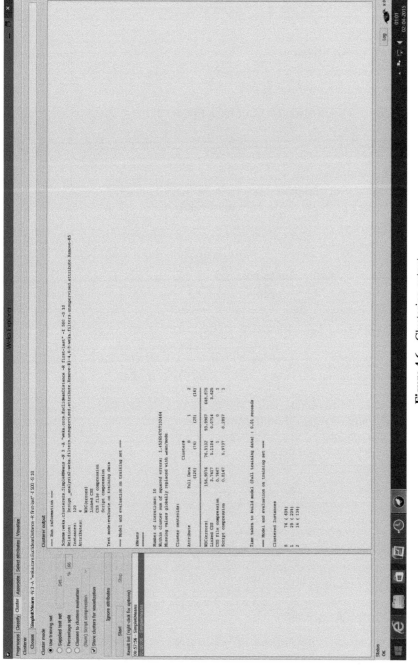

Figure 4.6 Clustering output.

Attributes: 4
    W3C(errors)
    Linked CSS
    CSS file compression
    Script compression
Test mode:evaluate on training data

=== Model and evaluation on training set ===

kMeans
======

Number of iterations: 10
Within cluster sum of squared errors: 1.452656787102464
Missing values globally replaced with mean/mode

Cluster centroids:

| | | Cluster# | | |
| Attribute | Full Data | 0 | 1 | 2 |
	(120)	(76)	(28)	(16)
W3C(errors)	156.9076	76.5132	93.9967	648.875
Linked CSS	3.7417	5.1184	0.0714	3.625
CSS file compression	0.7667	1	0	1
Script compression	0.8167	0.9737	0.2857	1

Time taken to build model (full training data): 0.01 seconds
=== Model and evaluation on training set ===
Clustered Instances
1   76 (63%)
2   28 (23%)
3   16 (13%)
****

It can be observed from the sum of squared errors, which is 1.452656787102464, that clusters were not that good. This value is due to missing values in some instances of the dataset. On an average, it was decided to accept the output as it is.

4.3.3 Cluster Analysis

Looking at cluster-1, the number of W3C errors are minimum compared with other clusters, i.e. 76.51 Linked CSS are also maximum compared with other clusters, which is again a plus point from design perspective. CSS file and script compression are also available as the value is 1 (1- indicates compression carried out and 0 indicate no compression is applied). Overall features of this cluster show that all the websites under this category have good design.

Cluster-2 has 28 websites under it. Looking at the W3C errors, they are quite less compared with Cluster-3 but do not have a good score for linked CSS. CSS files are not compressed, thus increasing the site loading time. Script compression is also carried out by very few websites as indicated by the score. Thus, websites under this cluster do not have a good design approach from SEO perspective.

It should be noted that 16% of the websites belong to cluster-3. It has maximum number of W3C errors, which is not good, making it less compatible with the international standards. There is presence of linked CSS and compressed CSS file and script, adding some plus points to this cluster. Compression value in the dataset is 1. All the websites belonging to this cluster have a moderate design approach.

4.3.4 Classification Through Clustering

Through clustering, websites are put into three different categories. Looking at the features of each category, a label can be assigned like cluster-1 that has all the websites, which are good from design perspective, and hence can be labeled as good and, similarly, cluster-2 as poor and cluster-3 as moderate. The arff file created as output of clustering is updated by replacing the cluster numbers with these labels. Following is the snapshot of the same.

```
@data
0,23,5,1,1,Good
1,103,9,1,1,Good
2,232,4,1,1,Good
3,77,14,1,1,Good
4,151,5,1,1,Good
5,12,0,0,0,Poor
6,19,16,1,1,Good
7,59,4,1,1,Good
8,205,2,1,1,Good
9,65,4,1,1,Good
10,504,4,1,1,Moderate
```

4.3.4.1 Classification via clustering using J48 algorithm

The updated arff file is used for classification. J48 classification method of Weka is used. Figure 4.7 shows the output of the classification technique applied. Websites are classified as Good, Moderate, and poor.

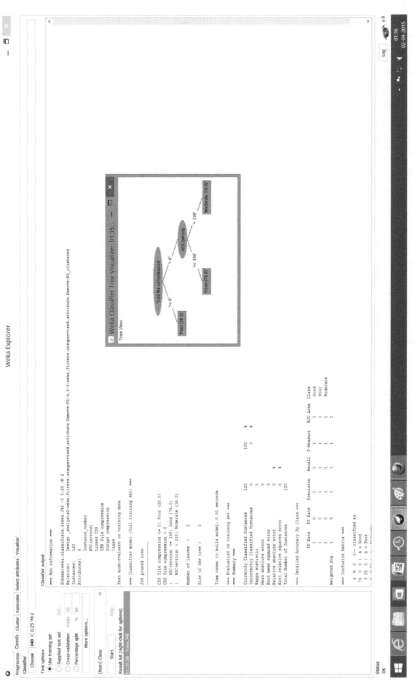

Figure 4.7   Classification based on design using J48 algorithm.

```
*****************************************************************
```

Scheme:weka.classifiers.trees.J48 -C 0.25 -M 2

Relation: Design_analysis2-weka.filters.unsupervised.attribute.Remove-R1-4,
6-8-weka.filters.unsupervised.attribute.Remove-R5_clustered

Instances: 120

Attributes: 6

    Instance_number
    W3C(errors)
    Linked CSS
    CSS file compression
    Script compression
    Class

Test mode:evaluate on training data

=== Classifier model (full training set) ===

J48 pruned tree

- - - - - - - - - - - - - - - -

CSS file compression $<= 0$: Poor (28.0)

CSS file compression > 0

| W3C(errors) $<= 338$: Good (76.0)

| W3C(errors) > 338: Moderate (16.0)

Number of Leaves: 3

Size of the tree: 5

Time taken to build model: 0.01 seconds

=== Evaluation on training set ===

=== Summary ===

Correctly Classified Instances	120	100%
Incorrectly Classified Instances	0	0%
Kappa statistic	1	
Mean absolute error	0	
Root mean squared error	0	
Relative absolute error	0%	
Root relative squared error	0%	
Total Number of Instances	120	

=== Detailed Accuracy By Class ===

TP Rate	FP Rate	Precision	Recall	F-Measure	ROC Area	Class
1	0	1	1	1	1	Good
1	0	1	1	1	1	Poor
1	0	1	1	1	1	Moderate
Weighted Avg.	1	0	1	1	1	1

```
=== Confusion Matrix ===
a    b    c    <- classified as
76   0    0  | a = Good
0    28   0  | b = Poor
0    0    16 | c = Moderate
```
****

Confusion matrix shows classification without errors, that is, all the instances are classified correctly. Figure 4.6 gives the graphical representation of classes. Out of 120 instances, 76 websites have good design approach and are hence SEO compatible. www.Myntra.com, www.Shoppersstop.com, www.Pantaloons.com, www.Redbus.in, etc., belong to this class; 16 websites have a moderate design approach as classified by the classifier. This class of websites needs to reduce their W3C errors to improve their SEO compatibility. www.naaptol.com, www.flipkart.com, www.amazon.in, etc., belong to this class. There are 28 websites out of 120 that need to redesign their websites, especially compression of CSS files and scripts. www.Zinghoppers.com, www.indianairlines.com, www.OnlineBuy99.com, etc., belong to this class.

4.3.4.2 Classification via clustering using RBFNetwork algorithm

A radial basis function network is an artificial neural network that uses radial basis functions as activation functions. The output of the network is a linear combination of radial basis functions of the inputs and neuron parameters.

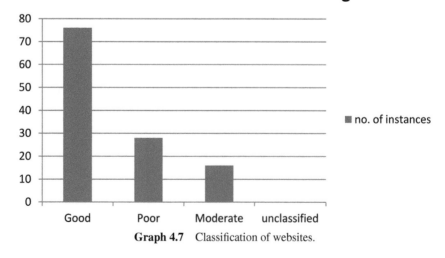

Graph 4.7   Classification of websites.

The dataset is loaded into Weka and then RBFNetwork classifier is fired on it. Figure 4.8 shows the output of executing RBFNetwork Weka classification algorithm on result.arff.

Confusion matrix shows that all instances of the dataset are classified correctly.
```
************************************************************
=== Confusion Matrix ===
a    b    c    <– classified as
76   0    0  |  a = Good
0    28   0  |  b = Poor
0    0    16 |  c = Moderate
************************************************************
```

Confusion matrix shows classification without errors, that is, all the instances are classified correctly. Graph 4.8 gives the graphical representation of classes. Out of 120 instances, 76 websites have good design approaches and are hence SEO compatible. www.Myntra.com, www.Shoppersstop.com, www.Pantaloons.com, www.Redbus.in, etc., belong to this class; 16 websites have moderate design approaches as classified by the classifier. This class of websites needs to reduce their W3C errors to improve their SEO compatibility. www.naaptol.com, www.flipkart.com, www.amazon.in, etc., belong to this class. There are 28 websites out of 120 that need to redesign their websites, especially compression of CSS files and scripts. www.Zinghoppers.com, www.indianairlines.com, www.OnlineBuy99.com, etc., belong to this class.

4.3.4.3 Classification via clustering using NaiveBayes algorithm

Naive Bayes classifiers are highly scalable, requiring a number of parameters linear to the number of variables (features/predictors) in a learning problem. The dataset is loaded into Weka and then, by using NaiveBayes classifier, it is classified into three distinct classes. Figure 4.9 shows the output of executing NaiveBayes Weka classification algorithm on result.arff.

Confusion matrix shows that except one, all instances of the dataset are classified correctly.
```
************************************************************
=== Confusion Matrix ===
a    b    c    <– classified as
75   0    1  |  a = Good
0    28   0  |  b = Poor
0    0    16 |  c = Moderate
************************************************************
```

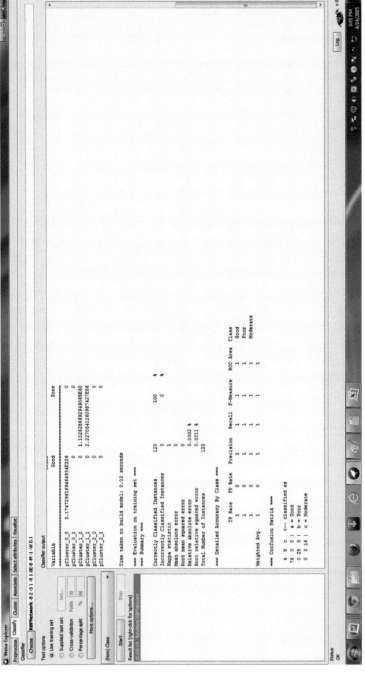

Figure 4.8   Classification based on design using RBFNetwork algorithm.

RBFNetwork classification based on Design

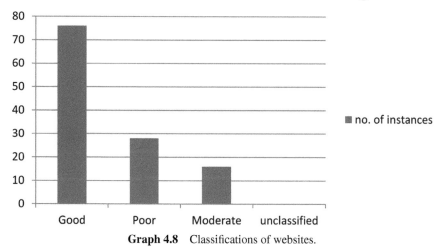

Graph 4.8   Classifications of websites.

Confusion matrix shows classification with one error. Graph 4.9 gives the graphical representation of classes. Out of 120 instances, 75 websites have good design approaches and are hence SEO compatible. www.Brandsn deals.com, www.Picsquare.com, www.PVRCinemas.com, www.Goair.in, www.SnapDeal.com, www.Jewelfunk.com, www.CromaRetail.com, etc., belong to this class; 16 websites have moderate design approaches as classified by the classifier. This class of websites need to reduce their W3C errors to improve their SEO compatibility. www.Babyoye.com, www.IndiaTimes.com, www.Junglee.com, www.Snapittoday.com, www.naaptol.com, etc., belong to this class. There are 28 websites out of 120 that need to redesign their websites, especially compression of CSS files and scripts.

4.3.4.4 Classification via clustering using SMO algorithm

In machine learning, support vector machines are supervised learning models with the associated learning algorithms that analyze data and recognize patterns, used for classification and regression analysis. It is used to classify linear or nonlinear data. Sequential Minimal Optimization (SMO), a fast method to solve huge quadratic programming problems, is widely used to speed up the training of the Support Vector Machines (SVMs). The dataset is loaded into Weka and then, by using SMO classifier, it is classified into three distinct classes. Figure 4.10 shows the output of executing SMO Weka classification algorithm on result.arff.

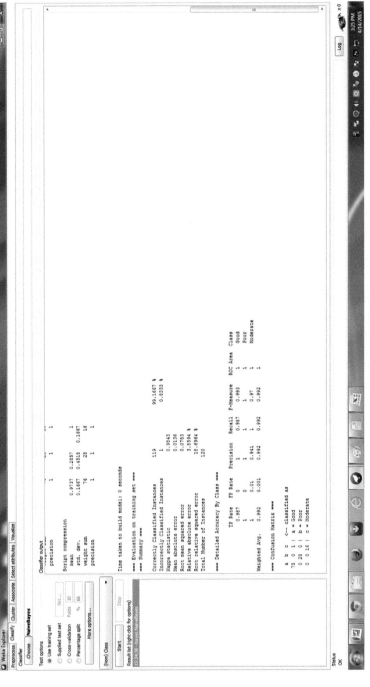

Figure 4.9   Classification based on design using NaiveBayes algorithm.

NaiveBayes classification based on Design

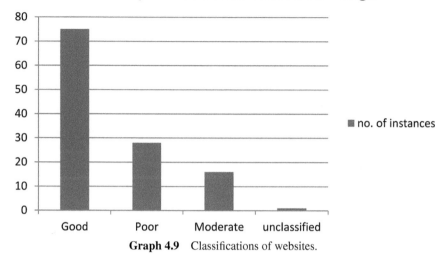

Graph 4.9 Classifications of websites.

Confusion matrix shows that 115 instances of the dataset are classified correctly and 5 instances are misclassified.

```
****************************************************************
=== Confusion Matrix ===
a   b    c    <- classified as
76  0    0  |  a = Good
0   28   0  |  b = Poor
5   0    11 |  c = Moderate
****************************************************************
```

Confusion matrix shows that 5 instances are misclassified. Graph 4.10 gives the graphical representation of classes. Out of 120 instances, 76 websites have good design approach and are hence SEO compatible. www.Brandsndeals.com, www.GlimGifts.com, www.Ixigo.com, www.Bagittoday.com, www.IndusCraft.com, etc., belong to this class; 11 websites have moderate design approach as classified by the classifier. This class of websites need to reduce their W3C errors to improve their SEO compatibility. www.Hoopos.com, www.Landmarkonthenet.com, www.Ebay.in, www.Junglee.com, www.naaptol.com, etc., belong to this class. There are 28 websites out of 120 that need to redesign their websites, especially compression of CSS files and scripts. www.Zinghoppers.com, www.Dealivore.com, www.RediffShopping.com, www.Onamgifts.com, www.Globus.com, etc., belong to this category.

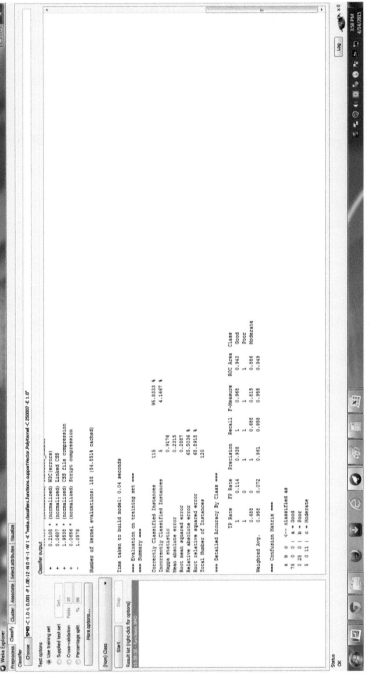

Figure 4.10   Classification based on design using SMO algorithm.

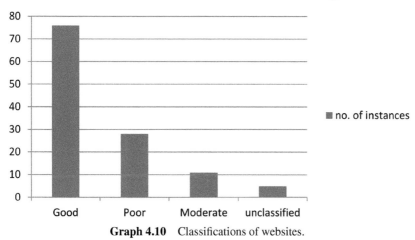

Graph 4.10 Classifications of websites.

4.3.4.5 Comparison of above classification algorithms

The above classification algorithms display the results in Tables 4.3 and 4.4, Graphs 4.11 and 4.12. Table 4.3 and Graph 4.11 display how 120 instances are classified correctly and incorrectly by the four classification algorithms supported by Weka. The time required to process the dataset is also given in seconds. From Table 4.3 and Graph 4.11, it is observed that the accuracies of the J48 and RBFNetwork algorithm are same. However, the time required to classify the instances are different. J48 took minimum time as compared with RBFNetwork algorithm. So, J48 algorithm has excellent performance with respect to accuracy and time.

From Table 4.4 and Graph 4.12, it is observed that kappa statistics values of J48 and RBFNetwork are same. But RBFNetwork has a Relative absolute error and a Root relative squared error. So, J48 is more suitable for classification of dataset of the business websites.

Table 4.3 Correctly and incorrectly classified instances

Sr. No.	Algorithm Used	Correctly Classified Instances	% of Correctly Classified Instance	Incorrectly Classified Instances	% of Incorrectly Classified Instance	Time in Sec.
1	J48	120	100	0	0	0
2	RBFNetwork	120	100	0	0	0.03
3	NaiveBayes	119	99.1667	1	0.8333	0
4	SMO	115	95.8333	5	4.1667	0.03

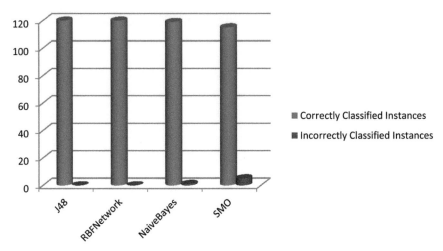

Graph 4.11   Correctly and incorrectly classified instances.

Table 4.4   Errors given by classification algorithms

Sr. No.	Algorithm Used	Kappa Statistic	Mean Absolute Error	Root Mean Squared Error	Relative Absolute Error	Root Relative Squared Error
1	J48	1	0	0	0	0
2	RBFNetwork	1	0	0	0.0002	0.0011
3	NaiveBayes	0.9843	0.0136	0.0783	3.8594	18.6964
4	SMO	0.9176	0.2315	0.2887	65.5036	68.8918

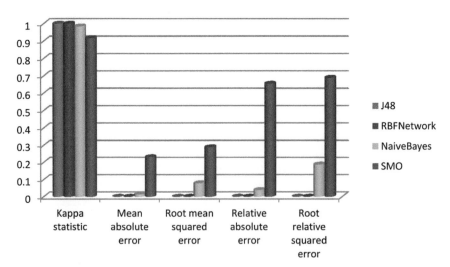

Graph 4.12   Errors given by classification algorithms.

4.4 Classification Based on Text

Text is an important part of webpage, as having only multimedia on the page is of no use unless and until information related to it is attached using text. It is believed that relevancy of the webpage by search engines involves parsing of the content of the webpage to collect the keywords. Thus, they directly as well as indirectly influence the SEO compatibility of the webpage.

The dataset for analysis of the webpage text is already created as mentioned in Chapter 3. It consists of 121 instances. Following are the attributes of the dataset stored as .csv file for analysis.

Page title: Page title should be unique and different from other websites and should be visible to users as well as search engines to increase SEO compatibility. It should include keywords and should be short and descriptive. The length of the page title should not be more than 55 characters to get good ranking.

Text-to-code ratio: It is an important attribute, as it is used by the search engines to find the relevancy of the webpage. The higher the ratio, the better the ranking by search engine, so it is an important attribute.

Titles: Under this, there are H1, H2, H3, H4, H5, and H6 sub heading. Among all the six, H1 tag is more important as it gives the information about the page and has more weightage compared with other tags as it helps in increasing search engine ranking, as search engines consider H1 tag for indexing.

Text styling: Three types of styling are used here for analysis, viz. Strong, U, and EM.

Strong tag helps in ranking the webpage, whereas U tag does not play any role. EM tag to some extent can help in ranking of the webpage.

Title coherence: It measures the number of keywords in the title. The ideal range is 2–4.

Keyword density: Keywords directly affect the search engine ranking and are important attributes. However, having many may cause keywords cumbersome in the webpage content, so optimum can be between 5 and 10.

4.4.1 Feature Selection

The above dataset is submitted to Weka for attribute selection. Figure 4.11 shows one of the outputs from the same.

After applying various attribute selection techniques and expert advice, the following attributes were selected for analysis: Page Title, Text-to-code ratio, H1, Strong, Title coherence, and keyword density.

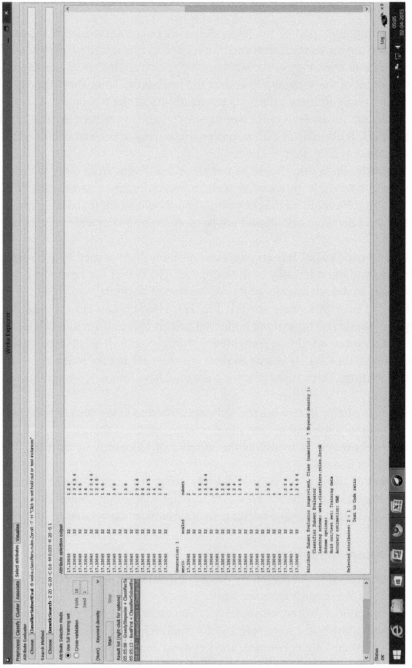

Figure 4.11   Selection of attributes.

4.4.2 Clustering

The above filtered dataset with six attributes is clustered into 3 categories using simple k-means clustering algorithm in Weka. Results show three good clusters with the sum of squared errors: 0.171416907883119 (Figure 4.12).

****

=== Run information ===

Scheme:weka.clusterers.SimpleKMeans -N 3 -A "weka.core.Euclidean Distance -R first-last" -I 500 -S 10

Relation: text_analysis-weka.filters.unsupervised.attribute.Remove-R1-2,6-10,12-weka.filters.unsupervised.attribute.Remove-R5

Instances: 121

Attributes: 6

    Page Title

    Text to Code ratio

    H1

    Strong

    Title coherence

    Keyword density

Test mode:evaluate on training data

=== Model and evaluation on training set ===

kMeans

======

Number of iterations: 5

Within cluster sum of squared errors: 0.171416907883119

Missing values globally replaced with mean/mode

Cluster centroids:

Attribute	Full Data	Cluster# 0	1	2
	(121)	(87)	(10)	(24)
Page Title	67.4463	83.2644	0	38.2083
Text to Code ratio	8.3159	5.9157	2.9	19.2729
H1	2.4463	3.2414	0	0.5833
Strong	6.8264	9.4138	0	0.2917
Title coherence	0.3169	0.1982	0	0.8796
Keyword density	52.1037	42.6482	97.334	67.5342

Figure 4.12 Clustering output.

Time taken to build model (full training data): 0.01 seconds
=== Model and evaluation on training set ===
Clustered Instances
1   87 (72%)   moderate
2   10 (8%)   bad
3   24 (20%)   good
****

4.4.3 Cluster Analysis

In Cluster-1, the page title has nearly 83 characters, which is not good as the optimum number is 55 or less than 55. The text-to-code ratio is also less, as the greater the ratio, the greater will be the page rank. But this ratio is better than cluster-2. H1 tag value is acceptable. Text style has good use of strong tag, which helps for SEO optimization and is better than other two clusters. Title coherence is moderate as compared with cluster-3. Keyword density as compared with cluster-3 is less. Overall, all the websites under this cluster have average performance from text attributes view.

Cluster-2 looks like outlier as most of the attribute values are null. It has more key word density and less text-to-code ratio. Overall, websites under this cluster need to focus on the text of the webpages.

Cluster-3 has 24 websites categorized under it. It has good page title score as it is within 55 characters making search engine friendly. The text-to-code ratio is also good compared with other clusters, but has less score over H1 heading and strong tag. Title coherence and keyword density are good.

4.4.4 Classification Through Clustering

The arff file which was produced as output of clustering is updated by changing the cluster no. to cluster label. Class labels are generated considering the features of each cluster from SEO perspective. So, cluster-0 is renamed as Moderate, cluster-1 as Bad, and cluster-3 as Good. Following is the sample arff file used for classification.

```
58,14,0.7,0,0,1,81.82,Good
59,66,1.13,0,4,0,50.29,Moderate
60,0,0,0,0,0,100,Bad
61,0,8.59,0,0,0,100,Bad
62,46,47.94,1,1,0.71,55.26,Good
63,131,0.27,61,0,0.12,42.55,Moderate
64,112,8.68,1,35,0.21,42.27,Moderate
```

4.4.4.1 Classification via clustering using J48 algorithm

This file is processed using Weka. J48 classifier is used for classification. Figure 4.13 shows the output of the classifier.

```
****************************************************************
= Run information ===
Scheme:weka.classifiers.trees.J48 -C 0.25 -M 2
Relation:   text_analysis-weka.filters.unsupervised.attribute.Remove-R1-2,6-
10,12-weka.filters.unsupervised.attribute.Remove-R5_clustered
Instances: 121
Attributes: 8
     Instance_number
     Page Title
     Text to Code ratio
     H1
     Strong
     Title coherence
     Keyword density
     Class

Test mode:evaluate on training data
=== Classifier model (full training set) ===
J48 pruned tree
- - - - - - - - - - - - - - - - - -
Title coherence <= 0.55
|Keyword density <= 81.82: Moderate (86.0/1.0)
|Keyword density > 81.82: Bad (10.0)
Title coherence > 0.55
| Strong <= 4: Good (23.0)
| Strong > 4: Moderate (2.0)

Number of Leaves: 4
Size of the tree: 7
Time taken to build model: 0.01 seconds
=== Evaluation on training set ===
=== Summary ===
Correctly Classified Instances        120        99.1736%
Incorrectly Classified Instances        1         0.8264%
Kappa statistic                    0.9809
Mean absolute error                0.0109
Root mean squared error            0.0738
```

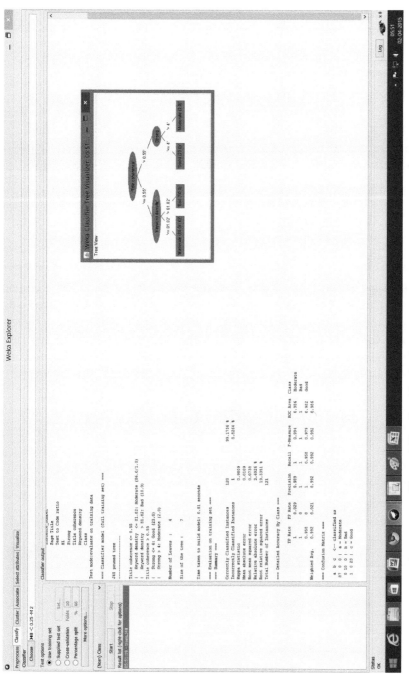

Figure 4.13   Classification based on text using J48 algorithm.

Relative absolute error   3.6926%
Root relative squared error   19.3351%
Total Number of Instances   121

=== Detailed Accuracy By Class ===

TP Rate	FP Rate	Precision	Recall	F-Measure	ROC Area	Class
1	0.029	0.989	1	0.994	0.986	Moderate
1	0	1	1	1	1	Bad
0.958	0	1	0.958	0.979	0.982	Good
Weighted Avg. 0.992	0.021	0.992	0.992	0.992	0.992	0.986

=== Confusion Matrix ===
```
 a  b   c   <- classified as
87  0   0 | a = Moderate
 0 10   0 | b = Bad
 1  0  23 | c = Good
```
****

Looking at the confusion matrix, the classifier has given 99% correct classes. Out of 121 instances, one is not correctly classified. Compared with classifications based on other criteria, it can be observed that website designers are not very serious about the text on the webpage as only 23 websites have good text on the webpage. www.A1books.co.in, www.Travelguru.com, www.Travelchacha.com, www.Letsbuy.com, etc., belong to this category.

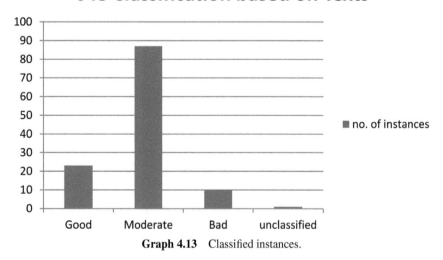

J48 Classification based on Texts

Graph 4.13   Classified instances.

There are nearly 72% websites, which are classified as moderate. This indicates that these websites are not that serious about the text on their webpages. www.flipkart.com, www.amazon.in, www.Pantaloons.com, www.Shoppersstop.com, etc., belong to this class.

Out of 121 websites, there are 10, whose website text is very poor making them very less compatible to SEO. Most of the data collected related to these websites have null entries. www.Hushbabies.com, www.indianairlines.com, www.OnlineBuy99.com, www.FutureBazaar.com, etc., are few websites from this class.

There is only one website, which is misclassified, thus giving an accuracy of 99%. Looking at all the other criteria classification results, this criterion is not given much importance by the website designers.

4.4.4.2 Classification via clustering using RBFNetwork algorithm

This file is processed using Weka. The RBFNetwork classifier is used for classification. Figure 4.14 shows the output of classifier.

```
**************************************************************
=== Confusion Matrix ===
a   b   c      <- classified as
86  0   1  |   a = Moderate
0   10  0  |   b = Bad
0   0   24 |   c = Good
**************************************************************
```

Looking at the confusion matrix, the classifier has given 99% correct classes. Out of 121 instances, one is not correctly classified. Compared with classifications based on other criteria, it can be observed that website designers are not very serious about the text on the webpage, as only 24 websites have good text on the webpage. www.Globus.com, www.Babyproducts.co.in, www.A1books.co.in, www.jetliteindia.co.in, www.Pepperfry.com, www.Lets buy.com, www.Zinghoppers.com, etc., belong to this category.

There are nearly 86 websites, which are classified as moderate. This indicates that these websites are not that serious about the text on their webpages. www.Elitify.com, www.ClearTrip.com, www.Mydala.com, www.Sportgenie.com, www.CromaRetail.com, etc., belong to this class.

There are 10 websites out of 121, whose website text is very poor making them very less compatible to SEO. Most of the data collected related to these websites have null entries. www.carkhana.com, www.Playgroundonline.com,

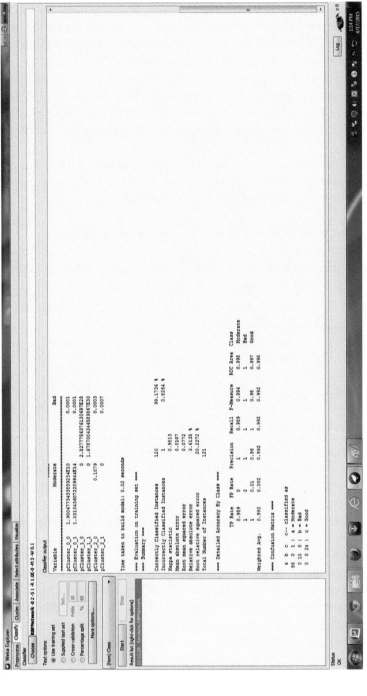

Figure 4.14 Classification based on text using RBFNetwork algorithm.

RBFNetwork Classification based on Texts

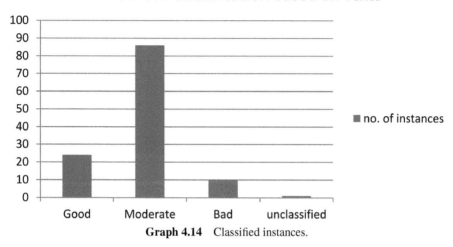

Graph 4.14 Classified instances.

www.Dealsonline4u.com, www.FutureBazaar.com, www.indianairlines.com, www.Hushbabies.com, etc., are few websites of this class.

There is only one website which is misclassified, thus giving an accuracy of 99%. Looking at all the other criteria classification results, this criterion is not given much importance by the website designers. The time required for classification of websites is 0.02 seconds.

4.4.4.3 Classification via clustering using NaiveBayes algorithm

This file is processed using Weka. NaiveBayes classifier is used for classification. Figure 4.15 shows the output of the classifier.

```
************************************************************
=== Confusion Matrix ===
a   b   c    <-- classified as
86  0   1  |  a = Moderate
0   10  0  |  b = Bad
0   0   24 |  c = Good
************************************************************
```

Looking at the confusion matrix, the classifier has given 99% correct classes. Out of 121 instances, one is not correctly classified. Compared with

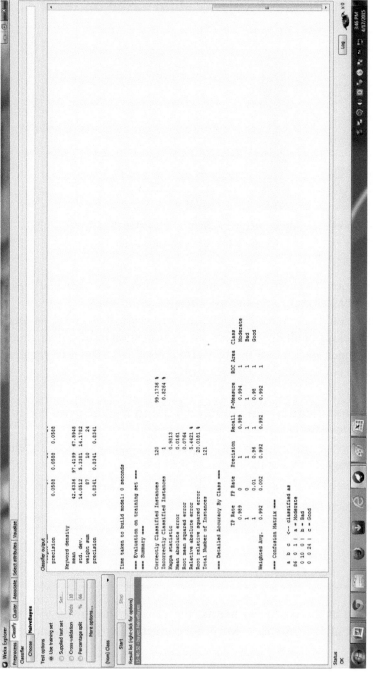

Figure 4.15   Classification based on text using NaiveBayes algorithm.

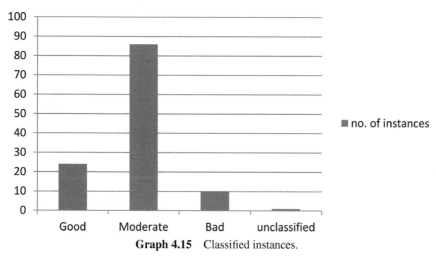

Graph 4.15 Classified instances.

classifications based on other criteria, it can be observed that website designers are not very serious about the text on the webpage, as only 24 websites have good text on the webpage. www.Zinghoppers.com, www.HealthShoppe.in, www.Dealivore.com, www.jetliteindia.co.in, www.Globus.com, etc.

There are nearly 86 websites, which are classified as moderate. This indicates that these websites are not that serious about the text on their webpages. www.Sunglassesindia.com, www.Infibeam.com, www.Travel Masti.com, www.Dealface.com, www.Perfume2Order.com, www.Croma Retail.com, etc., belong to this category.

There are 10 websites out of 121, whose website text is very poor making them very less compatible to SEO. Most of the data collected related to these websites have null entries. www.Hushbabies.com, www.indianairlines.com, www.BindaasBargain.com, www.Playgroundonline.com, www.carkhana.com, etc., are few websites of this class.

There is only one website which is misclassified, thus giving an accuracy of 99%. Looking at all the other criteria classification results, this criterion is not given much importance by the website designers.

4.4.4.4 Classification via clustering using SMO algorithm

This file is processed using Weka. SMO classifier is used for classification. Figure 4.16 shows the output of the classifier.

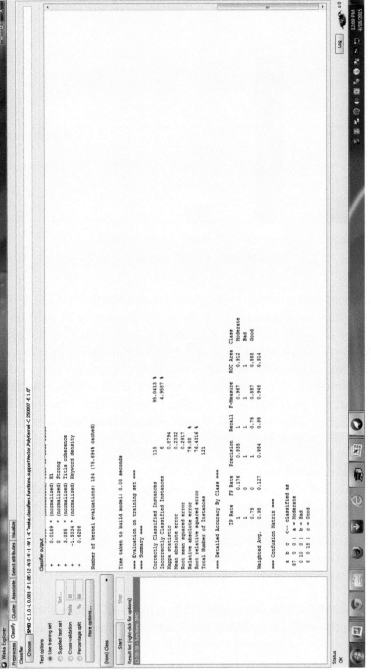

Figure 4.16 Classification based on text using SMO algorithm.

```
**************************************************************
=== Confusion Matrix ===
a   b   c    <– classified as
87  0   0  | a = Moderate
0   10  0  | b = Bad
6   0   18 | c = Good
**************************************************************
```

Looking at the confusion matrix, the classifier has given 95% correct classes. Out of 121 instances, six are not correctly classified. Compared with classifications based on other criteria, it can be observed that website designers are not very serious about the text on the webpage, as only 18 websites have good text on the webpage. www.Globus.com, www.BigFlix.com, www.Dealivore.com, www.Spectglasses.com, etc., belong to this category.

There are nearly 87 websites, which are classified as moderate. This indicates that these websites are not that serious about the text on their webpages. www.Picsquare.com, www.Yatra.com, www.Dealface.com, www.Housefull.co.in, etc., belong to this class.

There are 10 websites out of 121, whose website text is very poor making them very less compatible to SEO. Most of the data collected related to these websites have null entries. www.Hushbabies.com, www.OnlineBuy99.com, www.Playgroundonline.com, www.carkhana.com, etc., are few websites of this class.

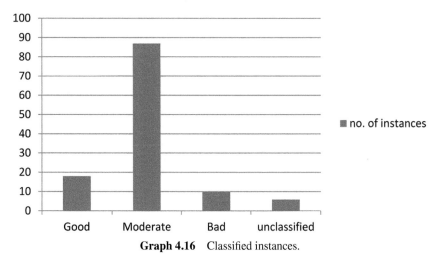

Graph 4.16 Classified instances.

There are six websites which are misclassified, thus giving an accuracy of 95%. Looking at all the other criteria classification results, this criterion is not given much importance by the website designers.

4.4.4.5 Comparison of above classification algorithms

The above classification algorithms display the results in the Tables 4.5 and 4.6, Graphs 4.17 and 4.18. Table 4.5 and Graph 4.17 display how 121 instances are classified correctly and incorrectly by the four classification algorithms supported by Weka. The time required to process the dataset is also given in seconds. From Table 4.5 and Graph 4.17, it is observed that the accuracies of the J48 and RBFNetwork, and NaïveBayes algorithms are same. However, the time required to classify the instances are different. J48 took minimum time as compared with RBFNetwork and NaiveBayes algorithms. So, J48 algorithm has excellent performance with respect to accuracy and time.

From Table 4.6 and Graph 4.18, it is observed that kappa statistics values of RBFNetwork and NaiveBayes are same. However, RBFNetwork took more time for classification of websites. J48 classification algorithm has 99% accuracy and it also took minimum time as compared with other classification algorithms. Therefore, J48 is more suitable for classification of dataset of the business websites.

Table 4.5   Correctly and incorrectly classified instances

Sr. No.	Algorithm Used	Correctly Classified Instances	% of Correctly Classified Instance	Incorrectly Classified Instances	% of Incorrectly Classified Instance	Time in Sec.
1	J48	120	99.1736	1	0.8264	0
2	RBFNetwork	120	99.1736	1	0.8264	0.02
3	NaiveBayes	120	99.1736	1	0.8264	0.01
4	SMO	115	95.0413	6	4.9587	0.08

Table 4.6   Errors given by classification algorithms

Sr. No.	Algorithm Used	Kappa Statistic	Mean Absolute Error	Root Mean Squared Error	Relative Absolute Error	Root Relative Squared Error
1	J48	0.9809	0.0109	0.0738	3.6926	19.3351
2	RBFNetwork	0.9813	0.0107	0.0772	3.6135	20.2272
3	NaiveBayes	0.9813	0.0161	0.0764	5.4421	20.0181
4	SMO	0.8794	0.2332	0.2917	79.08	76.4316

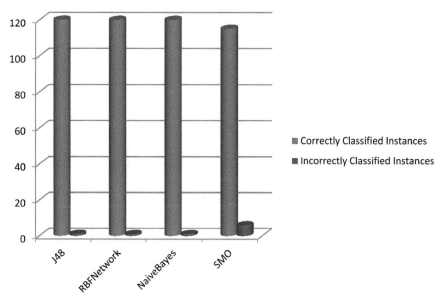

Graph 4.17   Correctly and incorrectly classified instances.

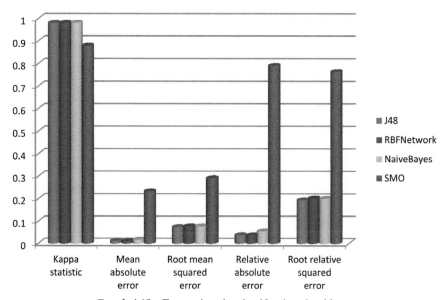

Graph 4.18   Errors given by classification algorithms.

4.5 Classification Based on Multimedia Content of Websites

Multimedia is an important tool to make the websites interesting and attractive. Multimedia may be in the form of images, videos, animation, High end graphics, or sound. Though it is interesting from user's point of view, it needs to be handled cautiously from developers point of view multimedia.

There are many reasons for this as the size of multimedia files is quite large, which will increase the page weight and ultimately increase the download time. If a webpage takes more time to download, the user will move away from it and change his navigation to other sites for faster accessibility. Developers from SEO point of view need to think about optimizing the multimedia files like .jpg extension that can be used for photographs and .png for rest of the image files. Optimum number of images per page also needs to be worked out as more images need more downtime and no images will be less interesting. Here is an attempt to classify websites depending on their multimedia content.

A dataset consisting of 118 instances has been created with all the attributes related to multimedia as discussed in Chapter 3. It has seven attributes as discussed below:

Number of images: The number of images is directly related to the loading time of a webpage. Loading time is an important factor, as if loading time is less, it is easy for the user to visit the webpage and ultimately easy for search engines for indexing. Hence, there should be optimum number of images per webpage.

Unreachable images: These are images that when clicked give error or goes in infinite loop. For a website to be SEO compatible, it should not have any unreachable images, as search engines will not wait for the website to load and reply access is not available.

Image Caching: Image caching can help in increasing the performance of the webpage and ultimately enhance the user experience. Caching will reduce the loading time and bandwidth consumption.

Percentage of images with alternative text (POIWAT): Having an alternate text to images is a plus point from search engine perspective, as it provides semantic description of images, and thus increases the search engine ranking of the webpage.

Percentage with correct alternative text (PWCAT): Having alternate text is not enough but having correct alternate text is important. If the alternate text is misleading, then it will decrease the rank of the page.

Frames and iFrames: iFrames are inline frames which allow embedding of another website, YouTube videos, PDF files, etc., in the code. This saves time but it is still controversial whether they are SEO friendly or not. iFrames can sometimes improve the SEO ranking, provided its contents are SEO friendly.

Favicon: These are small icons showing the first letter of a brand, logo of company, or a generic image. It is mainly made for humans rather than search engines, as brands and logos are for human recognition. This attribute has very minute effect on SEO, so it can be ignored in the analysis.

4.5.1 Feature Selection

The dataset with above-mentioned attributes was submitted to Weka for feature selection, so that only constructive attributes can be selected for further processing. Figure 4.17 is one of the attribute selection outputs using Weka.

After applying different attribute evaluators, the following five attributes were selected for analysis.

- Number of images
- Unreachable images
- Image Caching
- PWCAT
- Frames and iFrames.

However, as per the survey carried out, Frames and iFrames do not play any significant role in SEO compatibility, so this attribute is not considered in analysis.

4.5.2 Clustering

The filtered Multimedia dataset with only four attributes is used for processing. The dataset is clustered using Weka. Simple k-means clustering with 3 clusters is carried out. Output of clustering gave three good distinct clusters. Figure 4.18 shows the clustering result.

```
**************************************************************
=== Run information ===
Scheme:weka.clusterers.SimpleKMeans  -N  3  -A  "weka.core.Euclidean
Distance -R first-last" -I 500 -S 10
Relation:   Multimedia_analysis-weka.filters.unsupervised.attribute.Remove-
R1-2-weka.filters.unsupervised.attribute.Remove-R4,7-weka.filters.unsuper
vised.attribute.Remove-R5
```

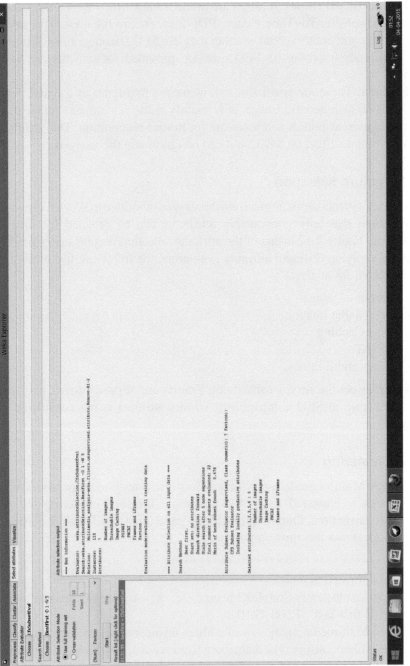

Figure 4.17 Attribute selection output.

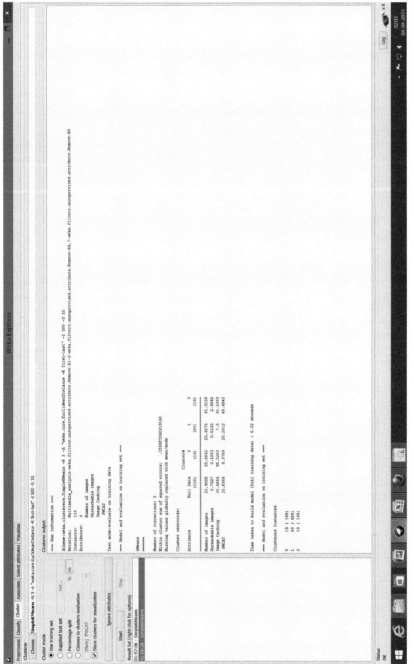

Figure 4.18 Clustering of multimedia dataset.

Instances: 118
Attributes: 4
    Number of images
    Unreachable images
    Image Caching
    PWCAT
Test mode:evaluate on training data

=== Model and evaluation on training set ===
kMeans
======
Number of iterations: 3
Within cluster sum of squared errors: .055987268619148
Missing values globally replaced with mean/mode
Cluster centroids:

Attribute	Full Data (118)	Cluster# 0 (19)	1 (80)	2 (19)
Number of images	25.9068	20.6842	23.4875	41.3158
Unreachable images	0.7627	0.1053	0.5125	2.6842
Image Caching	35.6864	98.5263	7.5	91.5263
PWCAT	25.8364	9.3789	20.2812	65.6842

Time taken to build model (full training data): 0.02 seconds
=== Model and evaluation on training set ===
Clustered Instances
1   19 (16%)   Good
2   80 (68%)   moderate
3   19 (16%)   poor
****

4.5.3 Cluster Analysis

Cluster-1 has 19 instances out of 118. Websites under this cluster have less number of images as compared with cluster-2 and cluster-3. It is a plus point, as having less number of images decreases the page weight and increases the speed of page load. Another good point here is that it has negligible number of unreachable images. Image caching is excellent but needs to improve on percentage with correct alternative text (PWCAT).

It was found that 68% of websites belong to cluster-2 with optimum number of images but little more than cluster-1. Unreachable images are more in this case than in cluster-1, but less than in cluster-3, so it has an average performance. It needs improvement in image caching, an important factor from user perspective. PWCAT ratio is better than that of cluster-1.

Cluster-3 has 16 websites under it. It has maximum number of images on the website, which reduces the speed of the website. Many of these images are unreachable, thus diverting the user navigation from the webpage. It has good percentage of image caching, but does not play a very important role as the number of images is also more. It has excellent PWCAT ratio compared with Cluster-1 and Cluster-2.

4.5.4 Classification Through Clustering

Using the features of each cluster, they are categorized as Good, Moderate and Poor in use of Multimedia in the website. These categories are used as class labels, and arff file of clustering result is updated as shown below:

```
15,6,1,100,15.4,Good
16,50,1,50,0,Moderate
17,50,1,50,0,Moderate
18,71,3,100,36.1,Poor
```

4.5.4.1 Classification via clustering using J48 algorithm

This updated file is used for classification using Weka. J48 classifier is used for classification. Figure 4.19 shows the classification based on multimedia using J48 algorithm.

Relation:Multimedia_analysis-weka.filters.unsupervised.attribute.Remove-R 1-2-weka.filters.unsupervised.attribute.Remove-R4,7-weka.filters.unsuper vised.attribute.Remove-R5_clustered-weka.filters.unsupervised.attribute. Remove-R1

Instances: 118

Attributes: 5

    Number of images
    Unreachable images
    Image Caching
    PWCAT
    Class

Figure 4.19   Classification based on multimedia using J48 algorithm.

****

Test mode: evaluate on training data
=== Classifier model (full training set) ===
J48 pruned tree
- - - - - - - - - - - - - - - - - -
Image Caching <= 52: Moderate (82.0/2.0)
Image Caching > 52
| PWCAT <= 34.5: Good (19.0)
| PWCAT > 34.5: Poor (17.0)
Number of Leaves: 3
Size of the tree: 5
Time taken to build model: 0 seconds
=== Evaluation on training set ===
=== Summary ===

Correctly Classified Instances	116	98.3051%
Incorrectly Classified Instances	2	1.6949%
Kappa statistic	0.9647	
Mean absolute error	0.022	
Root mean squared error	0.105	
Relative absolute error	6.7092%	
Root relative squared error	26.0161%	
Total Number of Instances	118	

=== Detailed Accuracy By Class ===

TP Rate	FP Rate	Precision	Recall	F-Measure	ROC Area	Class
1	0	1	1	1	1	Good
1	0.053	0.976	1	0.988	0.974	Moderate
0.895	0	1	0.895	0.944	0.957	Poor
Weighted Avg. 0.983	0.036		0.983	0.983	0.983	0.975

=== Confusion Matrix ===
a b c <- classified as
19 0 0 | a = Good
0 80 0 | b = Moderate
0 2 17 | c = Poor
****

Above graph is representation of confusion matrix. According to this, the classifier is 98% accurate as there are two instances which are misclassified.

97, www.HealthShoppe.in, 40,3,40,69.6,Moderate,Poor
100, www.Perfume2Order.com, 31,4,31,100,Moderate,Poor

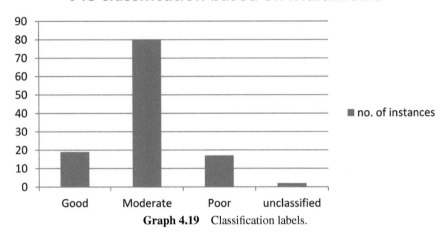

Graph 4.19   Classification labels.

Above two websites have been classified as poor but they belong to class moderate. The rest of the instances are classified correctly. Websites under class label Good have all the multimedia in optimum way. www.Pantaloons.com, www.Goair.in, www.Jabong.com, etc., belong to this class, whereas following websites are classified as moderate in multimedia performance: www.flipkart.com, www.amazon.in, www.Dell.co.in, etc.; 17 of the websites are labeled as Poor from multimedia perspective. www.Homeshop18.com, www.Yatra.com, www.Infibeam.com, etc.

4.5.4.2 Classification via clustering using RBFNetwork algorithm

This updated file is used for classification using Weka. RBFNetwork classifier is used for classification. Figure 4.20 displays the output of the RBFNetwork classifier.

```
************************************************************
=== Confusion Matrix ===
a   b   c    <- classified as
19  0   0  | a = Good
0   78  2  | b = Moderate
1   1   17 | c = Poor
************************************************************
```

Graph 4.20 is a representation of confusion matrix. According to this, the classifier is 96% accurate, as there are four instances which are misclassified.

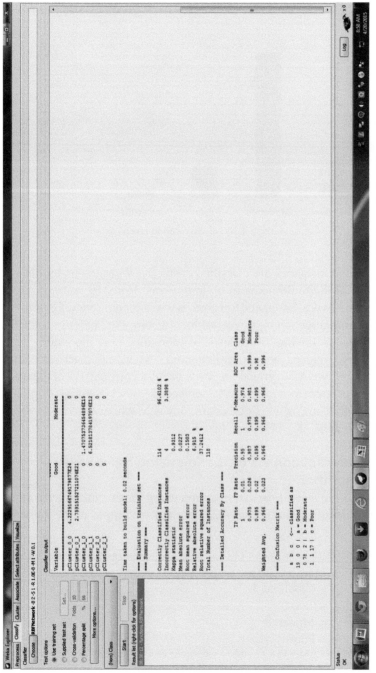

Figure 4.20 Classification based on multimedia using RBFNetwork algorithm.

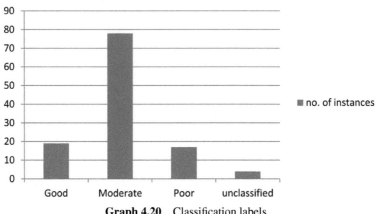

Graph 4.20   Classification labels.

The rest of the instances are classified correctly. Websites under class label Good have all the multimedia in optimum way. Here, 19 websites are classified as good. www.Shoppersstop.com, www.Via.com, www.Goodlife.com, www.PerfumesDirect.co.in, www.fashionara.com, etc., belong to this class, whereas the following websites are classified as moderate in multimedia performance: www.Orosilber.com, www.Landmarkonthenet.com, www.Grako.in, www.Mygrahak.in, www.naaptol.com, etc.; 17 of the websites are labeled as Poor from multimedia perspective. www.Sunglassesindia.com, www.Infibeam.com, www.Egully.com, www.Machpowertools.com, www. CromaRetail.com, etc., belong to this class.

4.5.4.3 Classification via clustering using NaiveBayes algorithm

This updated file is used for classification using Weka. NaiveBayes classifier is used for classification. Figure 4.21 displays the output of the NaiveBayes classifier.

```
*************************************************************
=== Confusion Matrix ===
a    b    c    <- classified as
18   0    1  |  a = Good
0    78   2  |  b = Moderate
0    2    17 |  c = Poor
*************************************************************
```

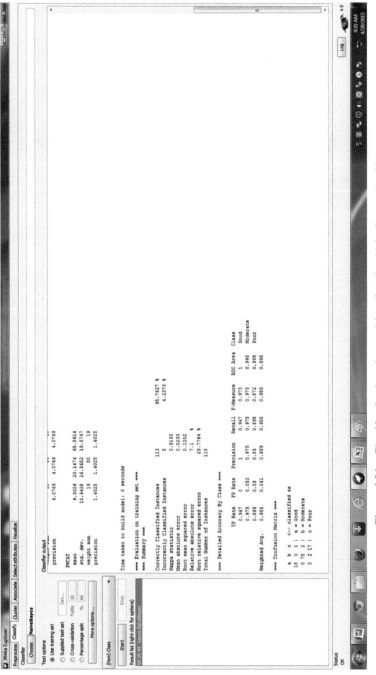

Figure 4.21 Classification based on multimedia using NaiveBayes algorithm.

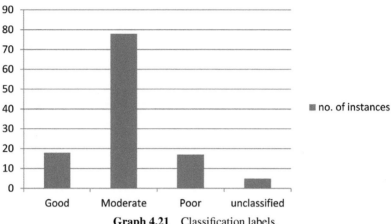

Graph 4.21   Classification labels.

Graph 4.21 is a representation of confusion matrix. According to this, the classifier is 95% accurate, as there are five instances which are misclassified. The rest of the instances are classified correctly. Websites under class label Good have all the multimedia in optimum way. Here, 18 websites are classified as good. www.Pantaloons.com, www.ClearTrip.com, www.SnapDeal.com, www.Getmecab.com, www.fashionara.com, etc., belong to this class, whereas the following websites are classified as moderate in multi-media performance: www.Orosilber.com, www.Redbus.in, www.Ebay.in, www.Playgroundonline.com, www.amazon.in, etc., 17 of the websites are labeled as Poor from multimedia perspective. www.Myntra.com, www.Yatra.com, www.Homeshop18.com, www.Dealface.com, www.Mach powertools.com, etc., belong to this class.

4.5.4.4 Classification via clustering using SMO algorithm
This updated file is used for classification using Weka. SMO classifier is used for classification. Figure 4.22 displays the output of the SMO classifier.

```
**************************************************************
=== Confusion Matrix ===
a    b    c    <- classified as
19   0    0  | a = Good
0    80   0  | b = Moderate
0    3    16 | c = Poor
**************************************************************
```

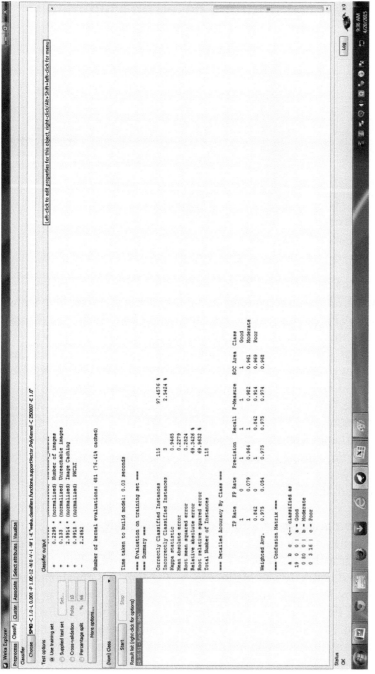

Figure 4.22 Classification based on multimedia using SMO algorithm.

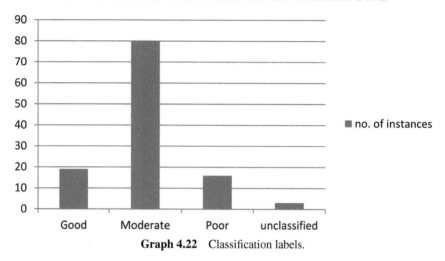

Graph 4.22   Classification labels.

Graph 4.22 is a representation of confusion matrix. According to this, the classifier is 97% accurate, as there are three instances which are misclassified. The rest of the instances are classified correctly. Websites under class label Good have all the multimedia in optimum way. Here, 19 websites are classifies as a good. www.Pantaloons.com, www.ClearTrip.com, www.Goodlife.com, www.PerfumesDirect.co.in, www.Getmecab.com, etc., belong to this class, whereas the following websites are classified as moderate in multimedia performance. www.Elitify.com, www.MakeMyTrip.com, www.Mallxs.com, www.Jpearls.com, www.Zinghoppers.com, etc.; 16 of the websites are labeled as Poor from multimedia perspective. www.Myntra.com, www.Yatra.com, www.Egully.com, www.LetsBuyProducts.com, www.Machpowertools.com, etc., belong to this class.

4.5.4.5 Comparison of above classification algorithm

The above classification algorithms display the results in Tables 4.7 and 4.8, Graphs 4.23, and 4.24. Table 4.7 and Graph 4.23 display how 118 instances are classified correctly and incorrectly by the four classification algorithms supported by Weka. The time required to process the dataset is also given in seconds. From Table 4.7 and Graph 4.23, it is observed that the accuracy of the J48 algorithms is more compared with other classification algorithms and the time required to classify the instances is 0.01 s. However,

Table 4.7 Correctly and incorrectly classified instances

Sr. No.	Algorithm Used	Correctly Classified Instances	% of Correctly Classified Instance	Incorrectly Classified Instances	% of incorrectly Classified Instance	Time in Sec.
1	J48	116	98.3051	2	1.6949	0.01
2	RBFNetwork	114	96.6102	4	3.3898	0.02
3	NaiveBayes	113	95.7627	5	4.2373	0
4	SMO	115	97.4576	3	2.5424	0.09

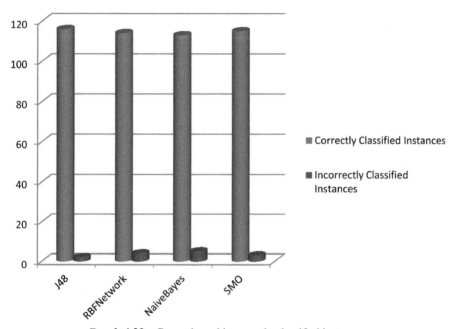

Graph 4.23 Correctly and incorrectly classified instances.

Table 4.8 Errors given by classification algorithms

Sr. No.	Algorithm Used	Kappa Statistic	Mean Absolute Error	Root Mean Squared Error	Relative Absolute Error	Root Relative Squared Error
1	J48	0.9647	0.022	0.105	6.7092	26.0161
2	RBFNetwork	0.9312	0.0227	0.1503	6.915	37.2412
3	NaiveBayes	0.9133	0.0233	0.1202	7.1	29.7764
4	SMO	0.9465	0.2279	0.2824	69.3426	69.9632

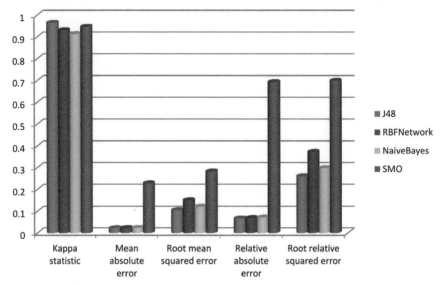

Graph 4.24   Errors given by classification algorithms.

the time required to classify the instances is minimum for NaiveBayes, but the accuracy is less compared with other classification algorithms. Therefore, J48 algorithm has better performance for classification of business websites.

From Table 4.8 and Graph 4.24, it is observed that kappa statistics value of J48 is very near to 1. J48 classification algorithm has 99% accuracy and it also took 0.01 s for classification of commercial websites. Therefore, J48 algorithm is more suitable for classification of dataset of the business websites.

4.6 Classification Based on Network Analysis of Webpage

Network analysis of a website deals with each webpage's network. A page network contains links, which are external and internal, indexing statistics and technical details for search engines. Links play an important role by providing certain comfort of navigation to users. When a webpage provides a link to a page which is not part of the host website, then it is treated as an external link, and if the link is to some section of the same webpage or another webpage of the same website, it is considered as an internal link. Search engines use link metrics to decide upon the importance they endow on each link. A website that has more number of links from other websites is considered as a trusted website and will rank higher in result pages of search engines. One more factor which is considered by search engines is the relevancy of the website passing the link. If you are a software developer and website passing link is a software company, then the link is considered relevant but if the linking website is of some paint company then it will be less relevant and thus will decrease the link juice.

There are two types of ranks associated with a website: domain rank and page rank. These ranks are important from search engine point of view. Google works with page rank, which is also known as page metric. Page rank increases if it has more links to it. Most of the website home pages have higher page metric compared with other pages associated with the website. Thus, the homepage will have more juicy content to link to. Similarly, contact us page will have only postal address and email addresses on it and will therefore have less page metric and low search ranking.

Here, network-based classification of websites will be carried out with emphasis on their ranking by search engines. Dataset is created using www.site-analyzer.com as mentioned in Chapter 3. Following are the attributes of dataset:

- **Number of links**: The total number of links associated with the website both internal and external.
- **Link juice**: It is associated with the relevancy of link passing website. If there is a webpage with many visitors, more browsing time, and maximum number of high-quality incoming links, then that webpage is performing well. If this webpage can pass link juice to a webpage which is not performing well than link juice of this webpage becomes positive, i.e. increases due to incoming link from good webpage.
- **Single links**: It is a ratio between single and duplicate links.
- **Links without underscore**: Search engines do not recognize underscore as word separator and hence will reduce the search index of the website.
- **Title attributes**: Measurement of the association between keywords and contents of the webpage.
- **Linking**: It is a percentage of no-follow links on the webpage.
- **Reliable links**: The ratio of the number of links available to search engine for crawling.

4.6.1 Feature Selection

All the attributes of the Network dataset may not be useful in classification, so attribute selection was carried out using select attribute menu of Weka. Different attribute evaluators were used. Figure 4.23 is the snapshot of the same.

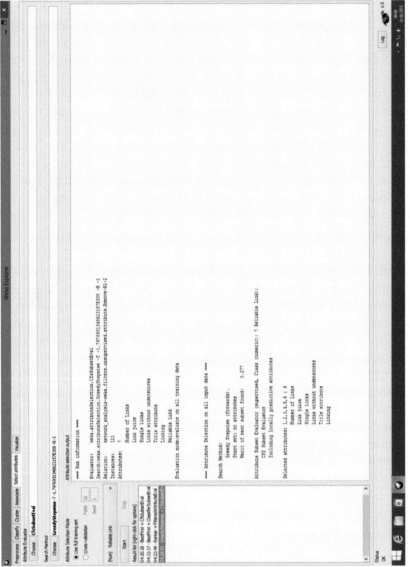

Figure 4.23   Feature selection output.

After applying different attribute selection tools, the following attributes were selected for analysis of dataset:

- Number of Links
- Link juice
- Single links
- Links without underscores
- Title attribute.

4.6.2 Clustering

Filtered network analysis dataset is clustered using Weka. Simple k-means clustering algorithm is executed on the dataset with 3 clusters. Figure 4.24 is the output of the clustered data in Weka.

```
**************************************************************
=== Run information ===
Relation:  network_analysis-weka.filters.unsupervised.attribute.Remove-R1-
2-weka.filters.unsupervised.attribute.Remove-R6-weka.filters.unsupervised.
attribute.Remove-R6
Instances: 111
Attributes: 5
    Number of Links
    Link juice
    Single links
    Links without underscores
    Title attribute
Test mode: evaluate on training data
=== Model and evaluation on training set ===
kMeans
======
Number of iterations: 4
Within cluster sum of squared errors: 0.0220784230053
Missing values globally replaced with mean/mode
```

Cluster centroids:

Attribute	Full Data	Cluster# 0	1	2
	(111)	(73)	(20)	(18)

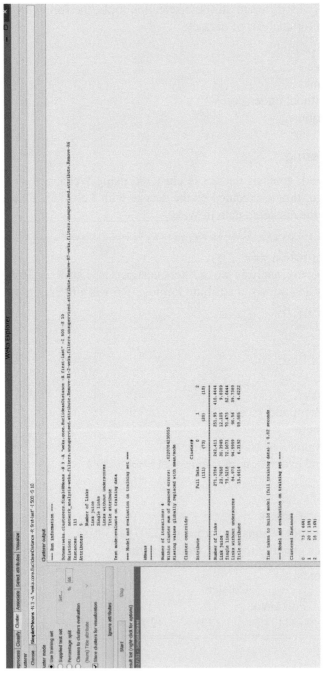

Figure 4.24   Clustering through Weka.

Number of Links	271.3784	242.411	251.95	410.4444
Link juice	23.7658	30.3945	12.105	9.8389
Single links	73.5218	72.1071	70.475	82.6444
Links without underscores	84.973	94.5959	90.56	39.7389
Title attribute	15.6414	6.3192	59.585	4.6222

Time taken to build model (full training data): 0 seconds
=== Model and evaluation on training set ===
Clustered Instances
0   73 (66%)
1   20 (18%)
2   18 (16%)
****

4.6.3 Observations

The aim of this classification is to classify websites on the basis of their availability to search engines, and two important attributes for search engines are link juice and links without underscore. Search engines do not use underscore to separate the words but use hyphen. The number of links may not play a crucial role here, as having more number of links may make the website design cumbersome. Title attribute does not carry much weight here, as search engines crawl through the contents of the websites for the keywords. Considering the above facts, cluster labels will be decided by analysis of intra-cluster data. For the convenience of readability, clusters are renamed as:

Clustered Instances
1   73 (66%)
2   20 (18%)
3   18 (16%)

A comparative study of all the three clusters can be carried out by using cluster centroids. Taking the first attribute, the number of links are as follows:

Number of Links: Cluster-1: 242.411
                    Cluster-2: 251.95
                    Cluster-3: 410.4444

The numbers of links are maximum in all websites under cluster-3. As per the Search Engine Optimization (SEO), drawback of having more links is the decrease in the internal page rank of links passes. If there are more links, the webpage will be like directory of links only without any information. Many users are visiting the websites for information and may not be interested in such webpages. If an information access needs many clicks, then it is not a user-friendly interface. As the analysis is of business websites, user-friendly interface is an important issue to be considered. Cluster-2 and Cluster-3 have moderate number of links associated with them. This attribute can be used in classification.

Link juice: Cluster-1: 30.3945
                 Cluster-2: 12.105
                 Cluster-3: 9.8389

It can be observed that Cluster-1 has maximum link juice followed by Cluster-2 and then Cluster-3. The more the link juice, the positive the link metric. Positive link factors that a link passes from one webpage to another are called link juice. How much link juice a link can pass from one page to another is determined by link metrics. Simple link between two webpages is sufficient to pass the link juice. In short, link juice of a webpage increases if the number of people provides links to that webpage. This link juice is associated with the SEO and will help in increasing ranking as well as more visitors to the website. Therefore, link juice plays an important role in search engine context and will be considered as one of the parameters in classification.

Single links: Cluster-1: 72.1071
                 Cluster-2: 70.475
                 Cluster-3: 82.6444

As mentioned above, a single link parameter gives the ratio between single and duplicate links on the webpage. This ratio should be good, as more single links help more to search engines. If there are duplicate links, search engines will not know which version to include in their indices and which version to rank for query results. IT is observed that there is not much distinction among the values of the attributes in all the three clusters, viz. cluster-1 and cluster-2 have almost same ratios and cluster-3 has a bit more than these two. As there is no much distinction among the values of the attribute, the single link parameter will not be considered for classification.

Links without underscores: Cluster-1: 94.5959
                              Cluster-2: 90.56
                              Cluster-3: 39.7389

It is advised to use hyphens (-) instead of underscores (_) in an URL. As Google uses hyphen as separator between words. Thus, using hyphens will make it more search engine friendly. It can be observed from clustering algorithm output that cluster-1 has maximum number of URLs without underscore, thus making it more search engine friendly and immediately followed by cluster-2 with a score of 90.56. Cluster-3 has poor score as only approximately 40% URLs are without underscore, making it less compatible to search engines. This attribute can play a role in classification, as the values among all the three clusters are quite distinct.

Title attribute: Cluster-1: 6.3192
                    Cluster-2: 59.585
                    Cluster-3: 4.6222

Title attribute gives extra information to the user, which appears when a mouse moves over an element, a tool tip appears showing text of the title. This attribute will not play any role in the classification of the websites, as this text is not used by the search engines either for retrieval or for ranking. Here, the focus of classification is on search engine, and hence this attribute will not be considered.

4.6.4 Classification Through Clustering

As per the above analysis, attributes which will be used for classification are the number of links, link juice, and links without underscore. Looking at the values of these attributes in all the three clusters, a label can be attached to each cluster.

Cluster-1 has moderate number of links which is good, good Link juice score: 30.39 and Links without underscores as 94.5959. This cluster can be labeled as more friendly with respect to search engines.

Cluster-2 also has moderate number of links little more than cluster-1, but poor link juice score: 12.105 and moderate number of links without underscore little less than cluster-1. Looking at all these values, this cluster can be labeled as moderate search engine friendly.

Cluster-3 has maximum number of links, making it less user friendly, link juice score 9.8389 is also very less, and links without underscore is 39.7389, making it less search engine friendly. Thus, it can be labeled as less search engine friendly. Here, result.arff is updated accordingly by replacing the cluster numbers with cluster labels.

4.6.4.1 Classification via clustering using J48 algorithm

The updated result.arff file is used as input to the classifier. Weka is used with J48 classifier. Figure 4.25 shows the output of Weka classifier.

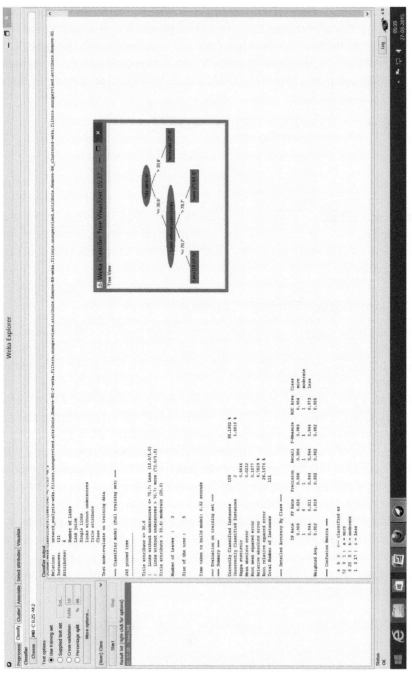

Figure 4.25   Classification based on networking using J48 algorithm.

****

Detailed output of classifier shows that classification is 98% accurate with only two instances incorrectly classified details are as follows:

Correctly Classified Instances   109   98.1982%
Incorrectly Classified Instances   2   1.8018%
Kappa statistic   0.9646
Mean absolute error   0.0232
Root mean squared error   0.1077
Relative absolute error   6.7829%
Root relative squared error   26.1474%
Total Number of Instances   111

=== Confusion Matrix ===
a   b   c   <– classified as
72 0   1 | a = more
0   20   0 | b = moderate
1   0   17 | c = Less
****

It can be observed from Graph 4.25 that, out of 111 websites, 72 are very compatible for search engines, thus having more visitors and good performance. Out of them, 20 are on average search engine friendly but need to increase their performance and 17 of the websites need to improve a lot

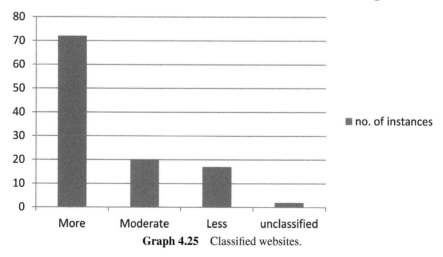

J48 Classification based on Networking

Graph 4.25   Classified websites.

to be ranked by the search engines. As the data collected is of commercial websites for business purpose, they need to be more search engine friendly to enhance their business.

4.6.4.2 Classification via clustering using RBFNetwork algorithm

The updated result.arff file is used as input to the classifier. Weka is used with RBFNetwork classifier. Figure 4.26 shows the output of Weka classifier.

****

Detailed output of classifier shows that classification is 97% accurate with only three instances incorrectly classified details are as follows:

Correctly Classified Instances	108	97.2973%
Incorrectly Classified Instances	3	2.7027%
Kappa statistic	0.9473	
Mean absolute error	0.0196	
Root mean squared error	0.1	
Relative absolute error	5.7319%	
Root relative squared error	24.2785%	
Total Number of Instances	111	

```
=== Confusion Matrix ===
 a   b   c    <- classified as
71   1   1 |  a = more
 1  19   0 |  b = moderate
 0   0  18 |  c = less
```
****

It can be observed from Graph 4.26 that, out of 111 websites, 71 are very compatible for search engines, thus having more visitors and good performance. Out of them, 19 are on average search engine friendly but need to increase their performance and 18 of the websites need to improve a lot to be ranked by the search engines. As the data collected is of commercial websites for business purpose, they need to be more search engine friendly to enhance their business.

4.6.4.3 Classification via clustering using NaiveBayes algorithm

The updated result.arff file is used as input to the classifier. Weka is used with NaiveBayes classifier. Figure 4.27 shows the output of Weka classifier.

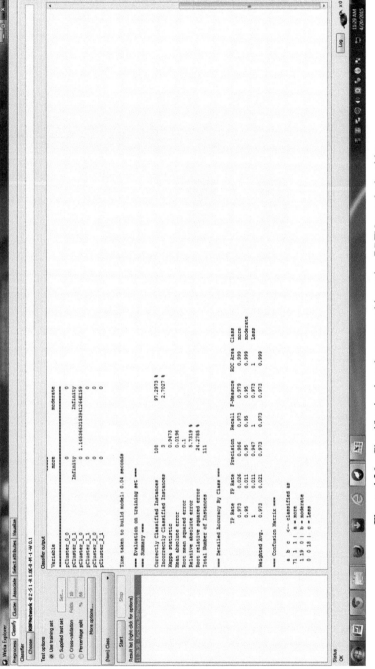

Figure 4.26 Classification based on networking using RBFNetwork algorithm.

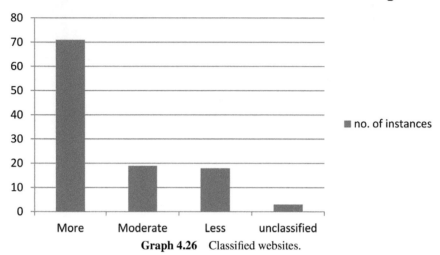

Graph 4.26   Classified websites.

```
****************************************************************
Correctly Classified Instances     99          89.1892%
Incorrectly Classified Instances   12          10.8108%
Kappa statistic                    0.8057
Mean absolute error                0.079
Root mean squared error            0.2149
Relative absolute error            23.108%
Root relative squared error        52.1722%
Total Number of Instances          111

=== Confusion Matrix ===
a    b    c    <– classified as
62   7    4  |  a = more
0    20   0  |  b = moderate
0    1    17 |  c = Less
****************************************************************
```

It can be observed from Graph 4.27 that, out of 111 websites, 62 are very compatible for search engines, thus having more visitors and good performance. Out of them, 20 are on average search engine friendly but need to increase their performance and 17 of the websites need to improve a lot to be ranked by the search engines. As the data collected is of commercial websites for business purpose, they need to be more search engine friendly to enhance their business.

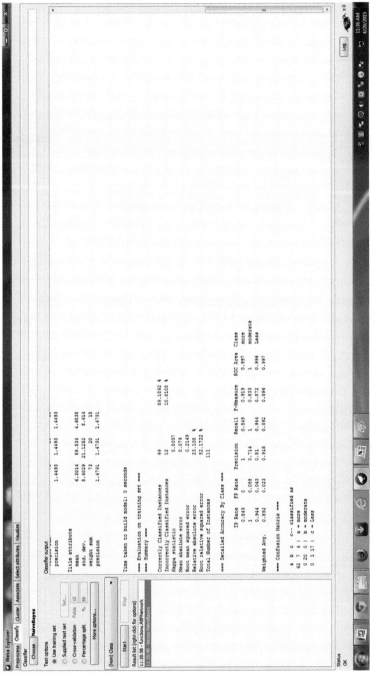

Figure 4.27 Classification based on networking using NaiveBayes algorithm.

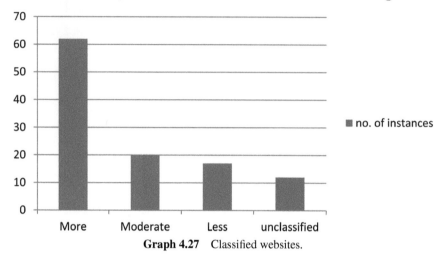

Graph 4.27   Classified websites.

4.6.4.4 Classification via clustering using SMO algorithm

The updated result.arff file is used as input to the classifier. Weka is used with SMO classifier. Figure 4.28 shows the output of Weka classifier.

```
****************************************************************
Correctly Classified Instances    103          92.7928%
Incorrectly Classified Instances  8            7.2072%
Kappa statistic                   0.8479
Mean absolute error               0.2402
Root mean squared error           0.3035
Relative absolute error           70.2619%
Root relative squared error       73.6858%
Total Number of Instances         111

=== Confusion Matrix ===
a   b    c    <- classified as
73  0    0  |  a = more
5   15   0  |  b = moderate
3   0    15 |  c = Less
****************************************************************
```

It can be observed from Graph 4.28 that, out of 111 websites, 73 are very compatible for search engines, thus having more visitors and good performance. Out of them, 15 are on average search engine friendly but need

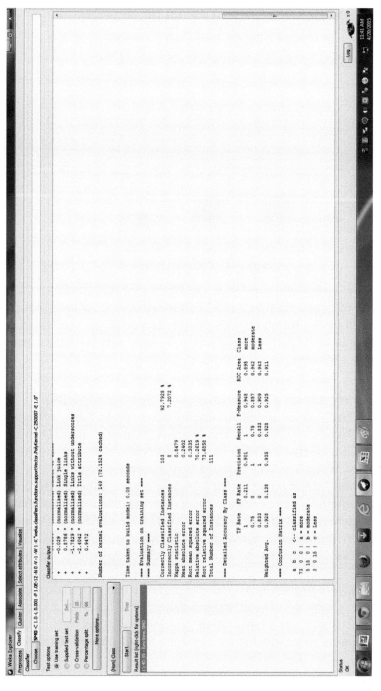

Figure 4.28 Classification based on networking using SMO algorithm.

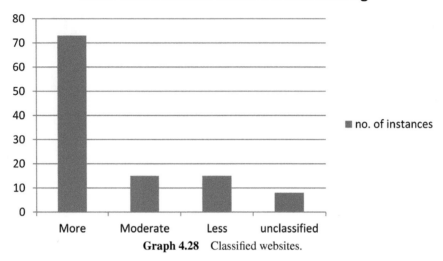

Graph 4.28   Classified websites.

to increase their performance and 15 of the websites need to improve a lot to be ranked by the search engines. As the data collected is of commercial websites for business purpose, they need to be more search engine friendly to enhance their business.

4.6.4.5 Comparison of the above classification algorithm

The above classification algorithms display the results in Tables 4.9 and 4.10, Graphs 4.29 and 4.30. Table 4.9 and Graph 4.29 display how 111 instances are classified correctly and incorrectly by the four classification algorithms supported by Weka. The time required to process the dataset is also given in seconds. From Table 4.9 and Graph 4.29, it is observed that the accuracy of the J48 algorithms is more compared with other classification algorithms and the time required to classify the instances is 0.02 s. However, the time required to classify the instances is minimum for NaiveBayes, but accuracy is 89%. Therefore, J48 algorithm has better performance for classification of business websites.

From Table 4.10 and Graph 4.30, it is observed that kappa statistics value of J48 is very near to 1. J48 classification algorithm has 98% accuracy and it took 0.02 second for classification of commercial websites. Therefore, J48 algorithm is more suitable for classification of dataset of the business websites.

Table 4.9 Correctly and incorrectly classified instances

Sr. No.	Algorithm Used	Correctly Classified Instances	% of Correctly Classified Instance	Incorrectly Classified Instances	% of Incorrectly Classified Instance	Time in Sec.
1	J48	109	98.1982	2	1.8018	0.02
2	RBFNetwork	108	97.2973	3	2.7027	0.04
3	NaiveBayes	99	89.1892	12	10.8108	0
4	SMO	103	92.7928	8	7.2072	0.08

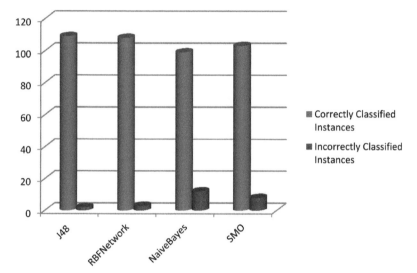

Graph 4.29 Correctly and incorrectly classified instances.

Table 4.10 Errors given by classification algorithms

Sr. No.	Algorithm Used	Kappa Statistic	Mean Absolute Error	Root Mean Squared Error	Relative Absolute Error	Root Relative Squared Error
1	J48	0.9646	0.0232	0.1077	6.7829	26.1474
2	RBFNetwork	0.9473	0.0196	0.1	5.7319	24.2785
3	NaiveBayes	0.8057	0.079	0.2149	23.108	52.1722
4	SMO	0.8479	0.2402	0.3035	70.2619	73.6858

4.7 Classification of Websites Using Overall Performance

In above sections, websites have been classified using different attributes like accessibility, design, text, multimedia, and networking. All these attributes play an important role in predicting the SEO compatibility of the websites.

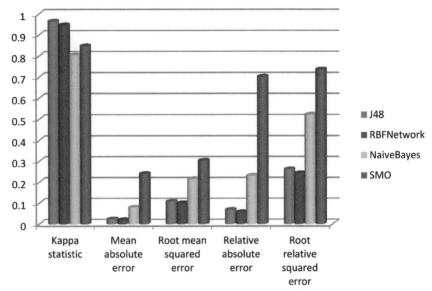

Graph 4.30 Errors given by classification algorithms.

This section will take the performance percentage of all these attributes to classify websites. The aim of classifying here is to find out the websites, which are good in all the attributes as above sections have classified the websites depending on individual attributes. Some websites may be performing well in accessibility and networking but needs improvement in text and design and is average in multimedia. Here, classification of websites will be done using all the attributes.

Dataset is created using the site-analyzer tool as mentioned in Chapter 3 with attributes like: global score, Name of the URL, Accessibility, Design, Text, Multimedia, and networking. Here, this dataset is not submitted to Weka for feature selection as all the attributes are vital and none can be eliminated. This overall.csv dataset consists of 123 instances.

4.7.1 Clustering

Clustering is carried out to find the labels for the class. Looking at the attributes of the dataset, it is decided to use global performance attribute for clustering as it consists of the average of all the attributes. A simple k-means clustering is carried out with three clusters as output. Figure 4.29 shows output of clustering using Weka.

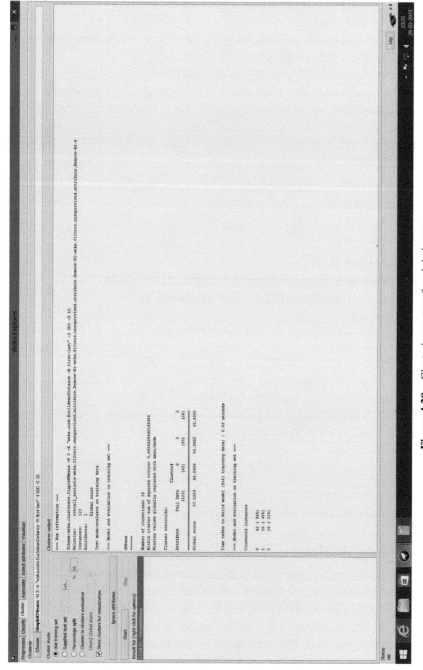

Figure 4.29 Clustering output for global score.

****

=== Run information ===

Scheme:weka.clusterers.SimpleKMeans -N 3 -A "weka.core.Euclidean
Distance -R first-last" -I 500 -S 10
Relation:overall_analysis-weka.filters.unsupervised.attribute.Remove-R1-we
ka.filters.unsupervised.attribute.Remove-R1-weka.filters.unsupervised.
attribute.Remove-R2-6
Instances: 123
Attributes: 1
    Global score

Test mode:evaluate on training data
=== Model and evaluation on training set ===

kMeans
======
Number of iterations: 19
Within cluster sum of squared errors: 0.8655625440165384
Missing values globally replaced with mean/mode
Cluster centroids:

		Cluster#		
Attribute	Full Data	0	1	2
	(123)	(42)	(55)	(26)
Global score	57.5289	64.0686	56.8062	48.4935

Time taken to build model (full training data): 0.02 seconds
=== Model and evaluation on training set ===
Clustered Instances
1   42 (34%)
2   55 (45%)
3   26 (21%)
****

4.7.2 Cluster Analysis

Three good clusters are generated with the sum of squared errors: 0.86. Let us look at the features of each cluster for analysis. Cluster-1 with 42 websites has an average global score of 64.068, which is the maximum among all the other clusters. Therefore, cluster-1 has all the websites which are SEO compatible and are performing well. Websites like www.Redbus.in, www.SnapDeal.com, www.Ebay.in, etc., fall in this cluster. This cluster is labeled as Best as per their performance.

There are 55 websites which belong to cluster-2 and have an average global score of 56.8062. These websites have good average score but need to improve still for more SEO compatibility and performance. Websites like www.Myntra.com, www.flipkart.com, www.naaptol.com, etc., fall in this cluster. Again looking at the overall performance of all the websites in this cluster, it is labeled as Good.

Cluster-3 has 26 websites with average performance of 48.49, which is neither best nor good but moderate. Websites in this cluster have to go for reverse engineering of their website developments and updates to improve the performance and increase SEO compatibility. www.Dealsonline4u.com, www.Shoppersstop.com, www.Pantaloons.com, etc., belong to this category. This cluster as per the performance is labeled as moderate.

4.7.3 Classification Via Clustering

Looking at the features of the clusters generated, a label is assigned to each cluster as discussed in the previous section. After getting the labels for the clusters overall.csv file is updated by adding an extra attribute class, as shown in Table 4.11.

4.7.3.1 Classification via clustering using J48 algorithm

This updated dataset is used for classification as class attribute is added to it. It is submitted to Weka for tree-based classification J48. Advantages of tree-based classifications have already been discussed in Section 4.2.4. Figure 4.30 shows the output of the classification.

```
************************************************************
=== Run information ===
Scheme:weka.classifiers.trees.J48 -C 0.25 -M 2
Relation:    overall_analysis-weka.filters.unsupervised.attribute.Remove-R1-
weka.filters.unsupervised.attribute.Remove-R1
Instances: 123
Attributes: 7
    Global score
    Accessibility
    Design
    Texts
    Multimedia
    Networking
    Class
```

Table 4.11 Updated overall.csv

Sr. No.	Name of the URL	Global Score	Accessibility	Design	Texts	Multimedia	Networking	Class
1	www.Sunglassesindia.com	59.23	73.72	62.02	46.67	55.88	50.82	good
2	www.Brandsndeals.com	56.01	71.29	52.73	52.15	38.23	58.19	good
3	www.Majorbrands.in	55.91	56.39	66.67	52.03	55.88	42.62	good
4	www.Pantaloons.com	47.9	46.42	87.43	29.69	35.88	14.75	moderate
5	www.Shoppersstop.com	51.31	64.56	65.46	57.13	24.12	27.87	moderate
6	www.Globus.com	59.61	60.81	54.64	66.36	74.71	49.18	good
7	www.Fetise.com	44.62	49.46	59.29	45.4	17.06	35.25	moderate
8	www.Orosilber.com	53.47	57.91	62.39	38.77	24.12	68.03	good
9	www.Violetbag.com	61.23	74.63	64.25	48.51	61.18	51.64	best
10	www.Myntra.com	55.11	53.11	70.81	29.24	61.18	52.46	good
11	www.Elitify.com	57.22	64.87	75.96	46.28	24.12	52.46	good
12	www.Picsquare.com	51.66	60.71	52.73	61.08	31.18	45.08	moderate
13	www.Yebhi.com	56.68	48.95	76.5	41.26	63.53	45.08	good

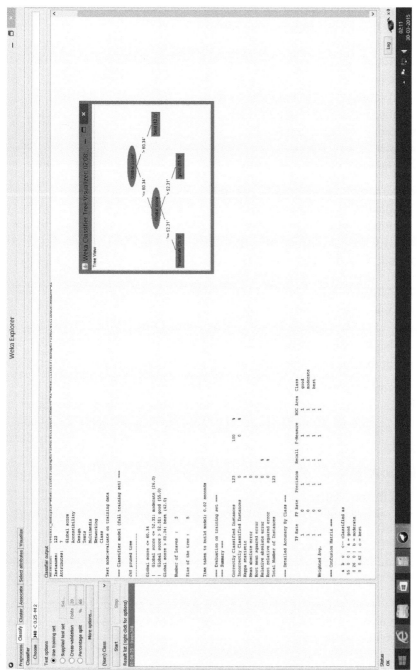

Figure 4.30 Classification based on overall using J48 algorithm.

Test mode:evaluate on training data
=== Classifier model (full training set) ===
J48 pruned tree

_____

Global score <= 60.34
| Global score <= 52.31: moderate (26.0)
| Global score > 52.31: good (55.0)
Global score > 60.34: best (42.0)
Number of Leaves: 3
Size of the tree: 5

Time taken to build model: 0.02 seconds
=== Evaluation on training set ===
=== Summary ===
Correctly Classified Instances   123 100%
Incorrectly Classified Instances   0   0%
Kappa statistic   1
Mean absolute error   0
Root mean squared error   0
Relative absolute error   0%
Root relative squared error   0%
Total Number of Instances   123

=== Detailed Accuracy By Class ===

TP Rate	FP Rate	Precision	Recall	F-Measure	ROC Area	Class
1	0	1	1	1	1	good
1	0	1	1	1	1	moderate
1	0	1	1	1	1	best
Weighted Avg. 1	0	1	1	1	1	

=== Confusion Matrix ===
a   b   c   <- classified as
55   0   0 | a = good
0   26   0 | b = moderate
0   0   42 | c = best
****

The classification carried out is 100% correct as per the confusion matrix all the instances are classified correctly. Graph 4.31 shows the graphical representation of the same.

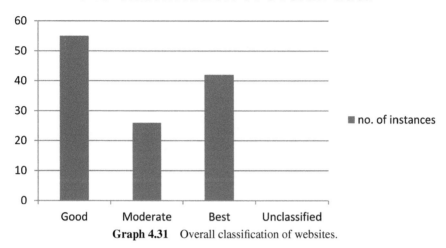

Graph 4.31 Overall classification of websites.

Taking into consideration all the attributes, the classifier gave good results and is matching with the real website experiences. www.flipkart.com, www.amazon.in, www.MakeMyTrip.com, etc., belong to this class.

There are nearly 42 websites which have the best performance considering all the aspects. This overall performance classification is dynamic, as if the values of any of the attribute change, the class of the website also changes.

4.7.3.2 Classification via clustering using RBFNetwork algorithm

This updated dataset is used for classification as class attribute is added to it. It is submitted to Weka for classification using RBFNetwork. Figure 4.31 shows the output of the classification.

The output of the classification of websites using RBFNetwork is given below:
```
****************************************************************
Correctly Classified Instances    122        99.187%
Incorrectly Classified Instances  1          0.813%
Kappa statistic                   0.9873
Mean absolute error               0.0072
Root mean squared error           0.0612
Relative absolute error           1.6903%
Root relative squared error       13.2694%
Total Number of Instances         123
```

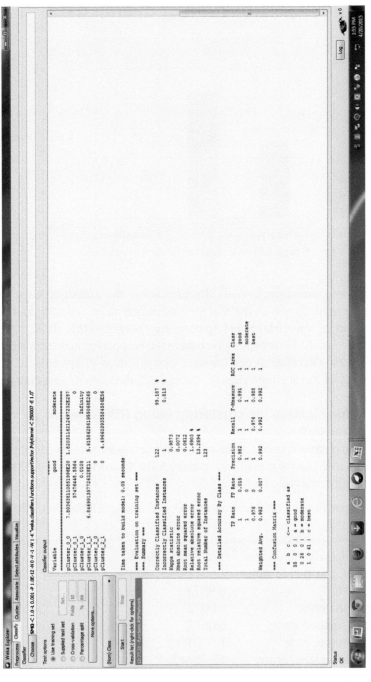

Figure 4.31   Classification based on overall using RBFNetwork algorithm.

=== Confusion Matrix ===
```
a   b    c    <- classified as
55  0    0  |  a = good
0   26   0  |  b = moderate
1   0    41 |  c = best
```
****

The classification accuracy is 99% and took 0.05 second for classification of business websites. Only one instance is classified incorrectly. Graph 4.32 shows the graphical representation of the same.

Taking into consideration all the attributes, the classifier gave good results and is matching with the real website experiences.

There are nearly 41 websites which have the best performance considering all the aspects. This overall performance classification is dynamic, as if the values of any of the attribute change, the class of the website also changes.

Website world is changing continuously due to its dynamic nature. Websites with informative contents, high accessibility, good loading speed with good user, and search engine interface are considered as SEO compatible. SEO has become an important criterion for business enhancement. It is recommended that not only business websites but also educational websites should be designed with SEO approach.

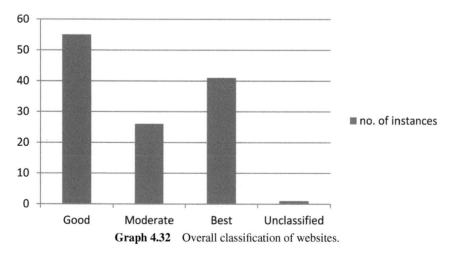

Graph 4.32 Overall classification of websites.

4.7.3.3 Classification via clustering using NaiveBayes algorithm

This updated dataset is used for classification as class attribute is added to it. It is submitted to Weka for classification using NaiveBayes. Figure 4.32 shows the output of the classification.

The output of the classification of websites using NaiveBayes is given below:

```
*************************************************************
Correctly Classified Instances    115        93.4959%
Incorrectly Classified Instances  8          6.5041%
Kappa statistic                   0.8983
Mean absolute error               0.08
Root mean squared error           0.1883
Relative absolute error           18.7566%
Root relative squared error       40.7976%
Total Number of Instances         123

=== Confusion Matrix ===
 a   b   c    <- classified as
50   2   3 |  a = good
 3  23   0 |  b = moderate
 0   0  42 |  c = best
*************************************************************
```

The classification accuracy is 93%. There are eight instances classified incorrectly. Graph 4.33 shows the graphical representation of the same.

Taking into consideration all the attributes, the classifier gave good results and is matching with the real website experiences.

There are nearly 42 websites which have the best performance considering all the aspects. This overall performance classification is dynamic, as if the values of any of the attribute change, the class of the website also changes.

Website world is changing continuously due to its dynamic nature. Websites with informative contents, high accessibility, good loading speed with good user, and search engine interface are considered as SEO compatible. SEO has become an important criterion for business enhancement. It is recommended that not only business websites but also educational websites should be designed with SEO approach.

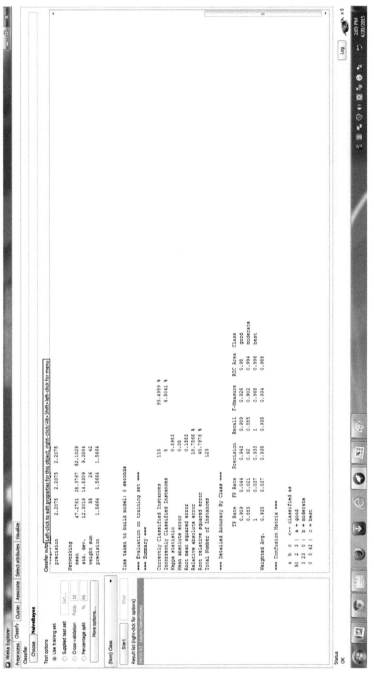

Figure 4.32 Classification based on overall using NaiveBayes algorithm.

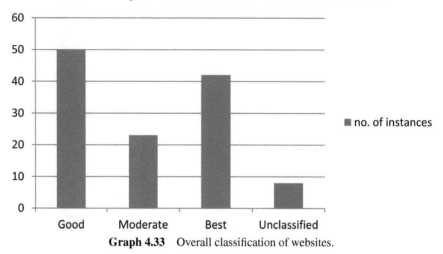

Graph 4.33   Overall classification of websites.

4.7.3.4 Classification via clustering using SMO algorithm

This updated dataset is used for classification as class attribute is added to it. It is submitted to Weka for classification using SMO. Figure 4.33 shows the output of the classification.

The output of the classification of websites using SMO is given below:
```
*************************************************************
Correctly Classified Instances    109          88.6179%
Incorrectly Classified Instances  14           11.3821%
Kappa statistic                   0.8164
Mean absolute error               0.2475
Root mean squared error           0.3152
Relative absolute error           58.0626%
Root relative squared error       68.3132%
Total Number of Instances         123

=== Confusion Matrix ===
a    b    c    <- classified as
54   0    1  |  a = good
8    18   0  |  b = moderate
5    0    37 |  c = best
*************************************************************
```

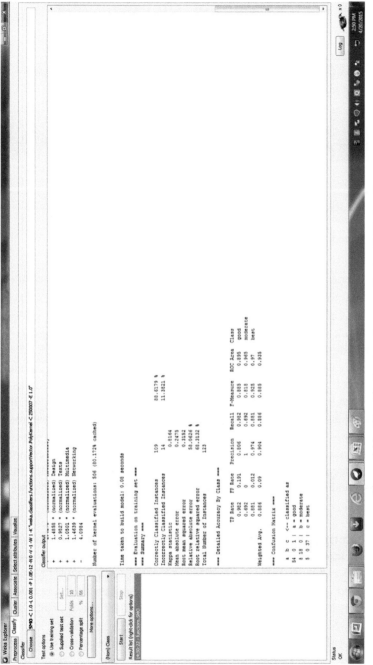

Figure 4.33 Classification based on overall using SMO algorithm.

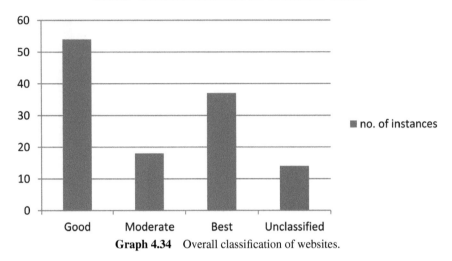

Graph 4.34   Overall classification of websites.

The classification accuracy is 88%. There are 14 instances classified incorrectly. Graph 4.34 shows the graphical representation of the same.

Taking into consideration all the attributes, the classifier gave good results and is matching with the real website experiences.

There are nearly 37 websites which have the best performance considering all the aspects. This overall performance classification is dynamic, as if the values of any of the attribute change, the class of the website also changes.

Website world is changing continuously due to its dynamic nature. Websites with informative contents, high accessibility, good loading speed with good user, and search engine interface are considered as SEO compatible. SEO has become an important criterion for business enhancement. It is recommended that not only business websites but also educational websites should be designed with SEO approach.

4.7.3.5 Comparison of the above classification algorithms

The above classification algorithms display the results in Tables 4.12 and 4.13, Graphs 4.35 and 4.36. Table 4.12 and Graph 4.35 displays how 123 instances are classified correctly and incorrectly by the four classification algorithms supported by Weka. The time required to process the dataset is also given in seconds. From Table 4.12 and Graph 4.35, it is observed that the accuracy of the J48 algorithm is high. However, the time periods required to classify the instances for J48 and NaiveBayes algorithms are same, i.e. zero. However,

Table 4.12 Correctly and incorrectly classified instances

Sr. No.	Algorithm Used	Correctly Classified Instances	% of Correctly Classified Instance	Incorrectly Classified Instances	% of Incorrectly Classified Instance	Time in Sec.
1	J48	123	100	0	0	0
2	RBFNetwork	122	99.187	1	0.813	0.05
3	NaiveBayes	115	93.4959	8	6.5041	0
4	SMO	109	88.6179	14	11.3821	0.9

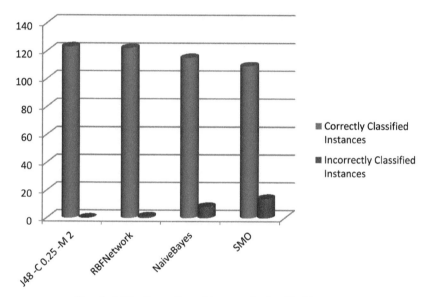

Graph 4.35 Correctly and incorrectly classified instances.

Table 4.13 Errors given by classification algorithms

Sr. No.	Algorithm Used	Kappa Statistic	Mean Absolute Error	Root Mean Squared Error	Relative Absolute Error	Root Relative Squared Error
1	J48	1	0	0	0	0
2	RBFNetwork	0.9873	0.0072	0.0612	1.6903	13.2694
3	NaiveBayes	0.8983	0.08	0.1883	18.7566	40.7976
4	SMO	0.8164	0.2475	0.3152	58.0626	68.3132

the accuracy of NaiveBayes is 93%. Therefore, J48 algorithm has excellent performance with respect to accuracy and time.

From Table 4.13 and Graph 4.36, it is observed that kappa statistics value of J48 is equal to one and all error terms are zero. That means accuracy is 100%.

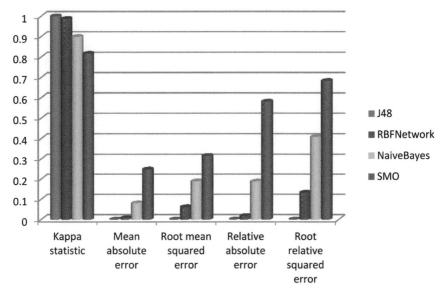

Graph 4.36   Errors given by classification algorithms.

Therefore, J48 is more suitable for classification of dataset of the business websites.

All the experiments carried out above for classification have proved that J48 is the most suitable classifier to classify websites based on their performance.

Website world is changing continuously due to its dynamic nature. Websites with informative contents, high accessibility, good loading speed with good user, and search engine interface are considered as SEO compatible. SEO has become an important criterion for business enhancement. It is recommended that not only business websites but also educational websites should be designed with SEO approach.

4.8 Results at a Glance and Conclusion

After exhaustive analysis, it is worthwhile to present the methodology and the results in the nutshell. As evidenced in this chapter, Website design plays a very important role in SEO. There are some parameters, which help for the proper design of the websites as per standards such as structure of the websites, link map, site map, user interface, browser, etc. Out of these, few features are used for data creation using the site-analyzer tool of 122 commercial websites.

Then, this dataset is preprocessed and filtered and then loaded into Weka for analysis. These freezed attributes such as W3C (errors), Linked CSS, CSS file compression, and Script compression are used for analysis. The k-means algorithm of clustering is used to categorize the commercial websites into three different clusters. The result obtained by using Weka is studied. Cluster-1 has 76 instances, Cluster-2 has 28 instances, and Cluster-3 has 16 instances. According their performance, the cluster numbers are replaced with good, poor, and moderate labels. The dataset is updated by using the new attribute as class. The updated dataset of commercial websites is loaded into Weka and classification.

The classification is carried on the dataset by using different classification algorithms, viz. J48 (Decision tree-based algorithm), RBFNetwork (neural network-based algorithm), NaiveBayes (statistical method), and SMO (Support vector machine-based algorithm). All the algorithms were executed using Weka. Out of these four classification algorithms, J48 and RBFNetwork classified all instances with the same accuracy, but RBFNetwork took more time as compared with J48, thus proving it as the good classifier compared with other algorithms.

The information of the content of the websites is possible only by using text. The site-analyzer tool displays the details of the 121 commercial websites, which are collected and stored in excel worksheet as a dataset. The dataset is preprocessed, filtered, and then loaded into Weka for further analysis. The freezed parameters such as page title, text-to-code ratio, H1, Strong, Title coherence, and Keyword density are used for analysis. In Weka, the dataset is partitioned into three different clusters by using k-means algorithm. The 121 instances are partitioned into three clusters. Out of 121, 87 are moderate, 10 are bad, and 24 are good. Then the new class label is added into the dataset for their performance in clustering. The updated dataset is loaded into Weka for classification. The classification is carried on the dataset by using different classification algorithms, viz. J48 (Decision tree-based algorithm), RBFNetwork (neural network-based algorithm), NaiveBayes (statistical method), and SMO (Support vector machine-based algorithm). All the algorithms were executed using Weka. Out of these four classification algorithms, J48, RBFNetwork, and NaiveBayes classified all instances with the same accuracy, but RBFNetwork and NaiveBayes took more time as compared with J48. Therefore, the performance of J48 is best.

Multimedia includes images, audios, videos, and high end graphics. The websites with multimedia contents take more time to download and increase the page weight. Owing to this, the user may lose his interest and turn to other websites for navigation. The site-analyzer tool is used to collect the dataset related to the parameters of multimedia. There are 118 commercial websites whose dataset is collected and stored. The collected data is preprocessed and filtered. The freezed attributes of the multimedia such as the number of links, unreachable images, image caching, PWCAT, and Frames and iFrames are loaded into Weka for analysis. The dataset is categorized into three different clusters. As per the result given by the k-means clustering algorithm, the cluster numbers are replaced with good, moderate, and poor labels. It is observed that, out of 118 clusters, 19 are good clusters, 80 are moderate clusters, and 19 are poor clusters. The dataset is updated by including the new attribute class into it. Then, the dataset is loaded into Weka for classification.

The classification is carried on the dataset by using different classification algorithms, viz. J48 (Decision tree-based algorithm), RBFNetwork (neural network-based algorithm), NaiveBayes (statistical method), and SMO (Support vector machine-based algorithm). All the algorithms were executed using Weka. Out of these four classification algorithms, performance of J48 was good.

Network analysis of a website is concerned with each webpage's network, which consists of internal and external links. If a website has more number of links from other websites, then its rank is high. There are two types of ranks such as domain rank and page rank associated with websites. In our research work, network-based classification of commercial websites is carried out with emphasis on their ranking by search engines. The dataset is collected and stored by using the site-analyzer tool and is also preprocessed and filtered. This preprocessed data is loaded into Weka and used for analysis. At the first, attributes are selected, which are number of links, link juice, single links, links without underscores, and title attribute. k-means algorithm is used to cluster the dataset into three different clusters. The results given by Weka are studied, and it is observed that, out of 111 instances, 73 form cluster-1, 20 form cluster-2, and 18 form cluster-3. As per the performance, cluster-1, cluster-2, and cluster-3 are replaced with more, moderate, and less labels. Then, the dataset is updated with new attribute class and loaded into Weka for classification. The classification is carried on the dataset by using different classification algorithms, viz. J48 (Decision tree-based algorithm), RBFNetwork (neural network-based algorithm), NaiveBayes (statistical method), and SMO (Support vector machine-based algorithm). All the algorithms were executed using Weka. Among all four algorithms, J48 gave good results.

As mentioned above, websites have been classified using different attributes like accessibility, design, texts, multimedia, and networking. These attributes helps predicting the SEO compatibility of the websites. The performance percentage of these attributes of the commercial websites is used to classify it. The overall performance of all commercial websites is tested by using these attributes values. The dataset is created using the site-analyzer tool with attributes such as global score, name of the URL, accessibility, design, texts, multimedia, and networking. This collected dataset is stored in excel worksheet, preprocessed and filtered, and then used for further analysis. Initially, it is loaded into Weka for analysis. All attributes are used for analysis purpose. The global score attributes give the global performance of the websites, so it is chosen for clustering. The simple k-means clustering algorithm is used to categorize the dataset of the commercial websites into three different clusters. As per the result given by Weka, the clusters are studied and cluster numbers are replaced with good, best, and moderated labels as per their performance. These cluster names are used to update the dataset by using new class label. The updated dataset is loaded into Weka for classification. The classification is carried on the dataset by using different classification algorithms, viz. J48 (Decision tree-based algorithm), RBFNetwork (neural network-based algorithm), NaiveBayes (statistical method), and SMO (Support vector machine-based algorithm). All the algorithms were executed using Weka. Out of these four classification algorithms, J48 and RBFNetwork classifiers have the same accuracy, but RBFNetwork took more time for the classification of datasets. Therefore, the performance of J48 is best.

4.9 Summary and Future Directions

The book showcases an effective methodology for classification of websites, which comes under the realm of web mining. There are many tools available freely on the Internet to analyze the websites for specific attributes. One of the tools used is the site-analyzer that could effectively generate the requisite dataset. Totally, 123 commercial websites were collected for analysis. These websites were processed using site-analyzer for analysis. This tool gave the detailed analysis of all the websites taking into consideration the Overall performance of the websites, user Accessibility, Design approach, Text on the webpage, Use of multimedia for attractive design and interface, and Link analysis using networking. Six datasets were created pertaining to each criterion. These generated datasets were pre-processed to be used with Weka for processing. Datasets were also filtered by removing some attributes having the same values for all the websites, as they do not play any role in distinguishing the websites. After pre-processing and filtering, these datasets were ready for classification.

Classification is a supervised method that needs some rules already generated to label the data. Here, an attempt was made to find the association between attributes to form a rule but did not give satisfactory results. Therefore, dataset attributes were reduced again by using attribute selection technique of Weka. Still, the results were not satisfactory as websites are dynamic and have diverse values for attributes. As difficulties were encountered due to the nature of the data in the attributes, it was decided to use cluster base approach. These data in the attributes were not suitable to directly decide the class label. The datasets were first clustered using simple k-means technique with two clusters. These generated clusters had high sum of squared errors, so the number of clusters were increased to three, which gave satisfactory output. Analysis of each cluster was performed to decide a label for these clusters. As clusters were three, it was decided to label them as Best, Good, and Moderate. In some datasets, we could not use the Label Best because of the values in the attributes, and so another set of labels for some attributes, Good, Moderate, and Poor, were used.

Once these cluster labels were ready, they were made as class label and the dataset was updated accordingly. This updated dataset was used for classification. Different classification algorithms were applied on the dataset to select the good classifier. Among all the algorithms, Decision tree technique gave good results due to its accuracy and time. Approximately, we could classify the datasets correctly except at few places, where there were one or two instances, which were misclassified as per confusion matrix.

The websites were classified approximately with some amazing results. There is not a single website, which is good in all the six attributes. There are some websites which are quite popular but have been classified as moderate in text-based classification. As all the websites were commercial, classification focused on SEO (Search Engine Optimal).

The book has been effective in carving a methodology in general, which can be applied to any domain. The salient features of such a methodology are

- Thorough literature review has been carried out.
- Domain used for research is business websites.
- Data is collected using the site-analyzer tool.
- Collected data is preprocessed and filtered to create the dataset for analysis.

- Accessibility, Design, Text, Multimedia, Networking, and Overall score were selected.
- Each attribute-related dataset was created using parameters of that attribute. A total of six datasets were created with approximately 123 instances.
- Datasets are filtered further by selecting the vital attributes, which are Search Engine optimized for processing using Weka attribute selection tool.
- Owing to the dynamic nature of dataset, it cannot be directly classified. Therefore, dataset is first clustered using k-means clustering in Weka.
- Clustered dataset is analyzed to assign labels for classification.
- Dataset with labels is classified by J48, RBFNetwork, NaiveBayes, and SMO techniques using Weka.
- A comparative analysis of all the classifiers was carried out.
- Amongst all the classification techniques, J48 (tree-based classifier) gave good results.
- Classification of the websites as Best, Good, Moderate, and bad performers from Search engine perspective.
- Research work carried out has good commercial applications in improving the website performance based on SEO.
- It also has good research applications as further analysis can be carried out using the same dataset with social media as focus.
- In academics, this can help in analyzing the student's interest in particular topics.

Application of the work carried out: As it is an era of e-commerce and e-marketing, the methodology described in the book can really help websites in finding their strengths and weaknesses. This analysis can help the business houses to increase their website SEO. Increased SEO will give good search engine indexing and ranking and will help in appearing on the first page of the search engine result. The authors are extending this work by including social networking aspects. Yet another direction of extension is applying SEO compatibility to different domains like Educational Institutes. A dynamic classification technique is being developed for real-time classification looking at the dynamic and ever changing nature of the websites.

Index

About the Authors

Dr. Vijaykumar S. Kumbhar is working as an Assistant Professor at Department of Computer Science, Shivaji University, Kolhapur. Spanning a long-standing career over a decade, Dr. Kumbhar is keen to mentor the students of two important postgraduate programs in Computer Science, namely Master in Computer Application (MCA) and Master of Science in Computer Science (MSc). Recently he pursued Ph.D. degree in Computer Science with title "Investigation of Classification Techniques for Domain Specific Data".

His other research interests include data mining, web mining, cloud computing, and information communication technology. He has completed a research project funded by University Grants Commission, Government of India. He has published good number of research papers in leading international and national journals in computer science. Dr. Kumbhar is also an active life member of Computer Society of India (CSI).

Dr. K. S. Oza is working as a faculty at Department of Computer Science, Shivaji University, Kolhapur. Spanning a long-standing career over a decade,

Dr. Oza is keen to mentor the students of two important postgraduate programs in Computer Science, namely Master in Computer Application (MCA) and Master of Science in Computer Science (MSc). Research being her passion, she enjoys guiding students for MPhil and Ph.D. in the research areas of data mining. Her other research interests include algorithms, data mining, cloud computing, and information communication technology. She has completed a research project funded by University Grants Commission, Government of India. She has published a good number of research papers in leading international journals in computer science. Her professional achievement has been widely appreciated by the computing community which has led to many international collaborations. The noteworthy among them are with the CQ University, Australia, under which an innovative project toward development of learning space for the computer science students is in progress. Dr. Oza is also an active life member of Computer Society of India (CSI).

Dr. R. K. Kamat is currently a Professor and Head, Department of Electronics, Shivaji University, Kolhapur. Prior to joining Shivaji University, he served in Goa University and on short-term deputation under various faculty improvement programs to Indian Institute of Science, Bangalore, and IIT Kanpur. He has successfully guided eleven students for Ph.D. in the research areas pertaining to Electronics and Computer science. His research interests include smart sensors, embedded systems, VLSI design, Information and Communication Technology (ICT) and Internet of things (IoT). He is the recipient of the Young Scientist Fellowship under the fast track scheme of Department of Science and Technology, Government of India, and extensively worked on Open Source Soft IP cores. He has published over 110 research papers, presented over 100 papers in conferences, and authored fifteen books. Dr. Kamat is currently Vice-President of IEEE India SSCS chapter and also a life member of Society of Advancement of Computing.